Artur Lugmayr, Samuli Niiranen, and Seppo Kalli

Digital Interactive TV and Metadata

Springer Science+Business Media, LLC

Engineering ONLINE LIBRARY

http://www.springer.de/engine/

Artur Lugmayr
Samuli Niiranen
Seppo Kalli

Digital Interactive TV and Metadata

Future Broadcast Multimedia

With 83 Illustrations

Springer

Artur Lugmayr
Samuli Niiranen
Seppo Kalli
Digital Media Institute
Tampere University of Technology
FIN-33101 Tampere
Finland

Library of Congress Cataloging-in-Publication Data
Lugmayr, Artur.
Digital interactive TV and metadata : Future broadcast multimedia
Artur Lugmayr, Samuli Niiranen, Seppo Kalli.
p. cm.
Includes bibliographical references and index.

1. Interactive television. 2. Metadata. 3. Television broadcasting—Technological innovations. I. Niiranen, Samuli. II. Kalli, Seppo. III. Title.
TK6679.3L84 2004
384.55—dc22 2004045616

Printed on acid-free paper.

ISBN 978-1-4419-1926-7 ISBN 978-1-4757-3953-4 (eBook)
DOI 10.1007/978-1-4757-3953-4

9 8 7 6 5 4 3 2 1

www.springer-ny.com

springeronline.com

To our families, friends and colleagues

Preface

Recent years have brought many changes to the world of mass media. The Internet and mobile communications technology have provided consumers with interactive digital services. Television is catching up with this trend through the digitalization process. Digital television is a hybrid platform combining elements from classical analog television and the Internet, providing modern multimedia services on a familiar platform. In short, digital TV is a gateway to the world of interactive digital media.

Digital TV brings consumers into the television service arena and offers them new degrees of freedom. However, as the service and multimedia content types diversify and the services and their content increase, television is facing many of the same challenges of complexity and information overflow faced by other digital media.

Metadata can handle the diverse services and content of digital TV efficiently and in a consumer-friendly way. Metadata means that the data are accompanied by other data which describe them. As data about data, metadata can provide an insight into syntactically and semantically complex data by distilling their essence to a set of simple descriptors. Metadata also helps to structure and manage information in diverse settings. The use of metadata in broadcast multimedia should not be restricted to being merely a tool for coping with the challenges of a complex networked multimedia environment. Instead, metadata offers new opportunities for the development of innovative services.

The research done by the *broadcasting multimedia group* at the Digital Media Institute (Institute of Signal Processing, Tampere University of Technology, Finland) has given us considerable experience and expertise with broadband multimedia. Our search for novel types of interactive service led to our applying metadata to digital TV. Metadata assists in bringing existing services to a new level and creating more advanced types.

Our research has focused on applying selected metadata standards to digital TV. The search for a unified solution for integrating metadata into the television service space showed MPEG-21 to be a good candidate solution for the creation, delivery and consumption of metadata-enabled services. The *digital broadcast item model (DBIM)* catalyzes the digital item methodology of MPEG-21 into a new converging concept for deploying metadata-based services in digital TV. The idea of the digital broadcast item model (DBIM) goes back to the year 2001, when the IEEE standard for the *learning object model (LOM)* was presented at a conference in Tampere, Finland. LOM introduced unified metadata structures for e-learning content. The authors realized that a similar structure was missing from the domain of digital TV. MPEG-21 was then identified as the natural starting point for the development of a unified metadata model for broadcast multimedia.

Our subsequent research work has focused on the development of a digital broadcast model, its accompanying service architecture and the services it can offer. The research work culminated in the establishment of an MPEG Ad-Hoc group "MPEG-21 in broadcasting" in 2003, chaired by the senior author of this book, to promote standardization within ISO/IEC.

This book describes how the use of the unified metadata model in digital broadcasting enhances traditional television service. Starting with a comprehensive overview of broadcast multimedia and related metadata, architectural design principles are presented for creating and using the digital TV platform services with a unified multimedia asset model within a metadata processing framework.

The digital broadcast item model represents a technical framework and a set of guidelines for managing services throughout the broadcast life-cycle. In short, it is a new converging concept for metadata in broadcasting. This is described in detail with emphasis on new innovative services and pathways that are likely to emerge over the next few years.

In the following there is a list with known trademarks mentioned in the book: IEEE ® is a registered trademark of The Institute of Electrical and Electronics Engineers, Incorporated; NOKIA ® is a registered trademark of Nokia Corporation; Sun ®, Java ® and all Java-based marks are trademarks or registered trademarks of Sun Microsystems, Inc. in the United States and other countries; DVB ® and MHP ® are registered trademarks of the DVB Project; SMPTE ® is a trademark of the Society of Motion Picture and Television; W3C ® is a trademark (registered in numerous countries) of the World Wide Web Consortium and marks of W3C are registered and held by its host institutions MIT, ERCIM, and Keio; CableLabs ® and OpenCable ® are trademarks of Cable Television Laboratories, Inc; the Bluetooth ® trademarks are owned by Bluetooth SIG, Inc. © Bluetooth SIG, Inc. 2004. ATSC, TV-Anytime, ISO/IEC, EBU and MPEG might be registered trademarks, but a relevant note on their Web-sites could not be found.

This book was written at the Digital Media Institute (Prof. Hannu Eskola, Director), Institute of Signal Processing (Prof. Moncef Gabbouj, Head) of the Tampere University of Technology, Finland. Our institute has provided a stimulating and open environment for the development of novel research ideas. We thank all our friends and colleagues for their discussions and companionship. Special credit goes to the members of the project teams of the broadcasting multimedia group, namely: Heikki Lamminen, Mathew Anurag Mailaparampil, Florina Tico, Mikko Oksanen, Perttu Rautavirta, Jussi Lyytinen, Heikki Mattila and Kirsi Keskiruusi.

Many thanks go to Prof. Frans Mäyrä of the Hypermedia Laboratory, Tampere University, Finland, for all his discussions about interactive narrative media and for his stimulating viewpoints. Credit also goes to Marie-Laure Ryan for many fruitful discussions about interactive narratives. We are indebted to Ismo Rakkolainen for discussions and for screenshots of his fog screen. Thanks go to Ville Holopainen for help with editing the numerous graphics of the book. Thanks also go to Johannes Messner at the department for scientific computing at the networking department at the Technical University of Linz for many years of LaTeX tips and tricks. We thank our publisher, Springer-Verlag New York, Inc., for their help and patience. Special thanks go to Margaret Mitchell. We would also like to thank Prof. Irek Defee and Prof. Reiner Creutzburg for their contributory discussions. Many thanks go to Hanna-Greta Puurtinen from eTampere for administrative help.

Many thanks also go to MPEG and all its marvelous members. We would like to say thanks to the MPEG Ad-Hoc group "MPEG-21 in broadcasting" members, especially to the co-chairs: Itaru Kaneko (Advanced Research Institute for Science and Engineering/RISE/Waseda University, Japan), Abdellatif Benjellountouimib (France Telecom, France) and Jong-Nam Kim (Korean Broadcasting System (KBS), Korea). To Prof. Andrew Perkis (NTNU, Norway) and Jan Bormans (IMEC, Belgium) go many thanks for bringing us closer to MPEG-21. Finally special thanks go to our families for all their love, friendship and support. The first author especially would like to thank Riina Pakarinen for her deep friendship.

The preparation of this book has been supported by a grant from the NOKIA Foundation.

More information, book errata, software and other novelties can be found on our Web-page at `http://www.digitalbroadcastitem.tv`.

Tampere,
March 22, 2004

Artur Lugmayr
Samuli Niiranen
Prof. Seppo Kalli

Contents

Part II Application

1

New Paradigms in Broadcast Multimedia

Broadcast multimedia is a common denominator for networked multimedia platforms that involve the use of a unidirectional broadcast channel to convey high-speed audio-visual services to consumers. The most common example of broadcast multimedia is the television. In the post-war era television has been the predominant consumer medium shaping basic expectations for electronic mass media. One such expectation is the passive role of the consumer in conventional television service utilization. As such, conventional broadcast television helped to establish the passive visual information absorption model of motion pictures for the home environment.

A basic technology trend in the evolution of networked multimedia platforms has been convergence towards digital information representation and conveyance. Digital television is the manifestation of this trend in the world of broadcast television. Through digitalization television is in a phase of metamorphosis evolving from a passive medium towards a fully interactive environment supporting a plethora of innovative service schemes. The possibilities of digital television extend far beyond the rigid service space of analog television by enabling the combination of audio-visual information to interactive services traditionally perceived as being foreign to television. Simply put, digital television represents a possibility of revolutionizing the way in which the general population perceives and deals with television.

As a related development, the emergence of the Internet as a consumer medium has led the way to more interactivity in the use of digital media by the general public. The Internet allows the consumer a maximal amount of freedom. The downside, however, is the chaotic nature of the Internet for novice users. Digital television represents a hybrid platform combining elements from both classical analog television and the Internet. It thus helps consumers to enjoy modern multimedia services on a fundamentally familiar platform. Digital TV is a gateway for the general public to the world of interactive digital media.

As a cautionary note, the world of television is changing rapidly and if the broadcast industry is not prepared for more active consumers they will face failure. A related phenomenon has already been experienced in the music industry, where digital media have provided a channel to distribute unauthorized copies of musical content on a world-wide scale. The availability of increased bandwidth and a comprehensive technology platform enabled this development. If the industry does not react to the changing needs of consumers used to the interactive Internet experience, television will lag behind the fundamentally interactive digital media.

1.1 Comparison of Classic Analog and Modern Digital TV

Figure 1.1 gives an abstract view of the differences between classic analog TV and modern digital TV.

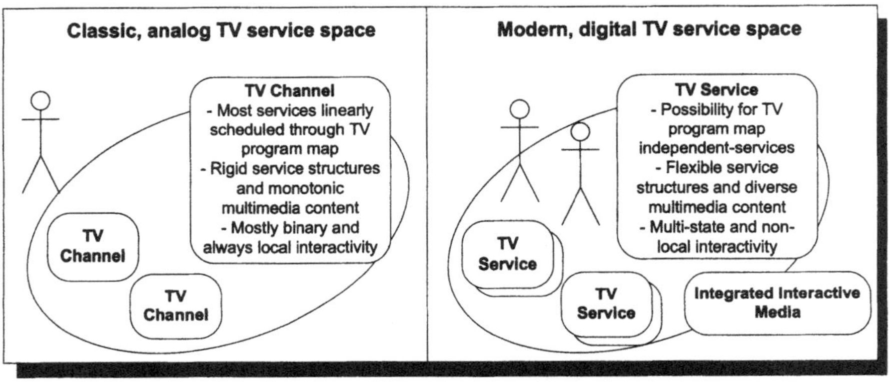

Fig. 1.1. An abstract view of the differences between classic analog TV and modern digital TV

Apart from the fundamental technological changes (e.g. the digital nature of the TV broadcast signal), digital TV revolutionizes the shape of the television service space.

As illustrated, the introduction of digital television extends the TV channel into TV service with radical changes in the structure and scope of services available. In analog TV most services (i.e. TV programs) are scheduled linearly according to a fixed TV program map. The common exception to this rule is the teletext service which can be accessed throughout the program map based on consumer initiative. Digital TV breaks away from this paradigm by allowing the provision and use of services independently of the broadcast TV program schedule. In classic TV, the service structure is very rigid and the types of available multimedia content are limited to basic A/V services and simple teletext content. Digital TV provides flexible service structures and a diversity of multimedia content types through the introduction of highly customizable value-added services.

Interaction in analog TV is typically limited to the selection of the TV channel viewed and to browsing of simple teletext service pages. Also, interactivity is always local: I.e. no electronically mediated information exchange between the consumer and another party is possible. Modern digitalized television brings more diversity into television interactivity. This is manifested in the flexible nature of value-added services and in the provision of integrated non local interaction facilities for information exchange with other parties.

A key concept related to availability of electronically mediated information exchange within the television experience is the introduction of communality. The integrated interaction facilities bring person-to-person interaction into television without the need for a separate medium to facilitate the non local information exchange. As a practical example, chat services existing in analog television that depend on the use of mobile phones to convey messages to the TV screen are replaced by services utilizing the native interaction capabilities of the digital TV consumer equipment.

As we have observed, digital TV brings consumers within the television service space and allows them a degree of freedom approaching that of Internet-type digital media. However, as the service and multimedia content types are diversified and the amount of services and content increases, television faces many of the same challenges as other digital media related to information overflow and loss of simplicity.

The use of metadata, or data about data, provides answers to how to handle the diverse services and contents of the new digital TV platform efficiently and in a consumer-friendly way. However, the use of metadata in broadcast multimedia should not be seen as just a tool to cope with the challenges inherent in a complex networked multimedia environment. Instead, it opens up new possibilities for the development of new innovative services.

1.2 First Thoughts about Metadata in Broadcast Multimedia

The concept of metadata conveys the idea of accompanying data with their descriptions. As data about data, metadata can provide an insight into syntactically and semantically complex data by distilling their essence into a set of simple descriptors. Metadata also helps to structure and manage information in diverse settings.

In broadcast multimedia metadata covers both the description of services and multimedia content. In short, metadata is used as a descriptive and structural framework for broadcast multimedia content and services.

Metadata integrates fully into the broadcasting value-chain with considerations for each step in the development of a digital TV broadcast service.

Figure 1.2 illustrates how basic digital audio and video signals, metadata extraction, metadata and services relate to each other. Metadata helps in the abstraction of the features of multimedia content and the development of new metadata-driven services. Segmentation of still images and visualization of image content within a value-added digital TV service (e.g. an EPG) is one example scenario for the utilization of metadata.

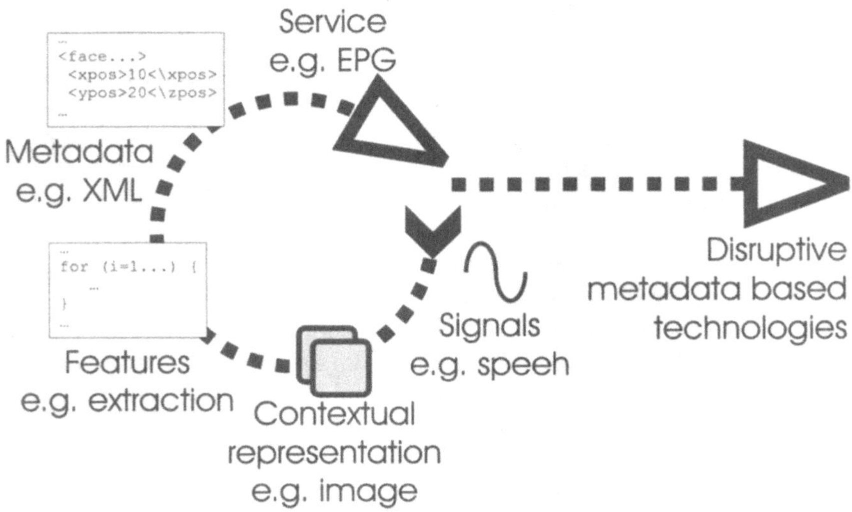

Fig. 1.2. An abstract lifecycle of metadata-driven digital TV services

Many different metadata standards relate to the digital broadcasting lifecycle. Each of these is defined by its system architecture, metadata definitions, work-flow or lifecycle model, typical characteristics and features, transmission formats, content representation and metadata representation.

The fundamental goal of this book is to elucidate how the different metadata standards converge into a unified approach enabling the development of new, metadata-driven broadcast multimedia services and value-chains.

1.3 Basic Definitions

Definition 1.1 (networked multimedia). *The concept of* networked multimedia *conveys the idea that multimedia content and services are provided in a distributed context utilizing the modern digital communication media.*

Definition 1.2 (broadcast multimedia). *The term* broadcast multimedia *is the common denominator for networked multimedia platforms that involve*

the use of a unidirectional broadcast channel to convey high-speed audio-visual services to consumers. The most common example of broadcast multimedia is television.

Definition 1.3 (metadata). *"Metadata is data about data... [and]... is information about a thing, apart from the thing itself" [19]. Metadata is "normally understood to mean structured data about resources that can be used to help support a wide range of operations. These might include, for example, resource description and discovery, the management of information resources and their long-term preservation" [46].*[1]

Definition 1.4 (multimedia asset). *A* multimedia asset *is any type of multimedia, either metadata definitions, or multimedia content asset or multimedia metadata asset. Multimedia metadata assets are a synonym for metadata definitions. Multimedia content assets relate to pure multimedia content. Atomic multimedia assets are useful measurable subunits of multimedia assets. Examples for multimedia assets are images, XML files and programs. Still frames of an MPEG-2 video stream are atomic multimedia assets.*

Definition 1.5 (multimedia content asset). *The term* multimedia content asset *defines several entities relevant for building multimedia services on any platform. They consist of arbitrary audio-visual multimedia elements, such as data, audio-visual information, presentations and animations. Multimedia content assets together with their description — multimedia metadata assets — build multimedia assets.*

Definition 1.6 (metadata definition language). *A* metadata definition language *is a domain independent description language to represent the structure, shape and types of arbitrary entities or their features of instantiated metadata documents. XML schema is an example of this type of metadata definition.*

Definition 1.7 (metadata definition instance). *A* metadata definition instance *represents an instantiated metadata definition language, which is correctly validated and syntax conformant. Where a metadata definition language describes inter document dependencies, a metadata definition instance holds concrete values. An XML document is an example of a metadata definition instance.*

Definition 1.8 (metadata definition). *A* metadata definition *is any type of metadata, thus it is either a metadata definition language or a metadata definition instance. It describes any type or form of metadata.*

Definition 1.9 (rigid metadata). Rigid metadata *refers to application specific, non customizable standardization of metadata definitions. These are e.g.*

[1] From our viewpoint the definitions stated by Ned Batchelder and Michael Day describe the purpose and aims of metadata in the best way.

the metadata definitions currently standardized in broadcast multimedia containing TV-program descriptions (DVB-SI, MPEG-2 PSI).

Definition 1.10 (granular metadata). Granular metadata *refers to metadata with a higher degree of flexibility and customizability than rigid metadata. Often it represents an extension of rigid metadata. MPEG-7 standards describing multimedia content assets are an example of granular metadata.*

1.4 Structure of the Book

The book consists of two parts: *Theory* and *Application.* Part One, Theory, includes the following chapters.

- Chapter 2 reviews the basic broadcast multimedia and digital TV standards and defines a starting point for the development of the metadata-based methodologies in digital TV. Introduction of MPEG-2, Digital Video Broadcasting (DVB), Multimedia Home Platform (MHP) and the basic digital TV value-chain are the highlights of the chapter.
- Chapter 3 reviews the existing metadata standards related to digital broadcasting and first considers their convergence into a unified metadata framework. Among the covered metadata standards are those defined by the W3C, MPEG and SMPTE.
- Chapter 4 defines the digital broadcast item model (DBIM) based on MPEG-21. It represents a unified metadata-based methodology including a technical framework and a set of guidelines for managing metadata-driven services throughout the digital broadcasting value-chain.
- Chapter 5 reviews general metadata-enabled service architectures and system views.

Part Two, Application, includes the following chapters.

- Chapter 6 focuses on new innovative, metadata-driven concepts and paradigms in digital TV. A consumer model and a narrative model are presented with consideration for the business models of metadata-driven digital TV service provision.
- Chapter 7 gives an overview of protection mechanisms for multimedia assets, transactions and digital rights management systems. It is a supportive technology for services.
- Chapter 8 includes an overview of metadata in digital production and delivery. Various service types and technologies related to the topic are covered.
- Chapter 9 addresses intelligent presentation and intelligent interaction models in digital TV.
- Chapter 10 focuses on how consumer profiling and personalization can utilize a metadata-based approach for customizing services.

- Chapter 11 concentrates on digital TV as a potential solution for the intelligent multimedia home. In ambient TV the consumer is embedded into an intelligent service space.
- Chapter 12 briefly describes other potential services.
- Chapter 13 maps the road ahead for digital TV and metadata-driven services.

Part I

Theory

2

World of Digital Interactive TV

2.1 Broadcast Multimedia

Broadcast multimedia are discussed here in the context of digital television standards. Europe, North America and Japan have established the basic standardization efforts for digital television. The basic technology for realizing digitalization in broadcast services is the MPEG-2 standard. MPEG-2 defines the basic concepts for the provision of digital A/V services in all the digital TV standardization efforts. MPEG-2 defines a signal compression, packetization and multiplexing standard for digital audio and video as well as rudimentary mechanisms for conveying other data in a digital A/V stream.

2.1.1 MPEG-2

The systems part of the ISO/IEC 13181 standard, which defines MPEG-2, addresses the combination of single or multiple *elementary streams (ES)* of video, audio and other data into streams suitable for storage or transmission. The specification basically enforces syntactic and semantic rules for systems coding and establishes two stream definitions for packet-oriented multiplexes: *transport stream (TS)* and *program stream (PS)*. Both stream definitions have a specific set of application areas. The most important application area of the TS is digital TV [1].

The basic packet-oriented multiplexing approach for video, audio and other data defined in MPEG-2 is illustrated in Fig. 2.1. Digital video and audio data is encoded into compressed MPEG-2 video and audio elementary streams as defined in the ISO/IEC 13181-2 and 13181-3, respectively. These elementary streams are packetized to produce video and audio *packetized elementary stream (PES)* packets. Private data is fed directly to the PES packetizer. PES packet streams are multiplexed to single transport or program streams depending on the application.

The program stream is designed for use in relatively error-free environments such as DVDs. On the other hand, the transport stream is a stream

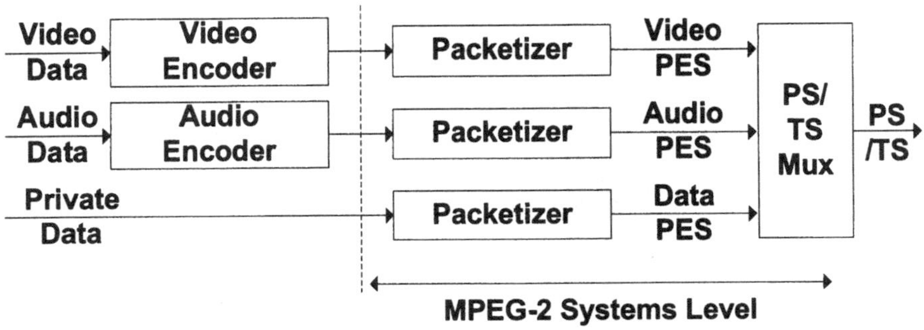

Fig. 2.1. MPEG-2 video, audio and data multiplex

definition tailored for communicating or storing streams of PES packets in error-prone environments. The errors may appear as bit value errors and loss of packets. The 188 byte transport stream packets begin with a 4 byte prefix containing a 13 bit *packet ID (PID)*. The PID identifies, via *program-specific information (PSI)* tables, the elementary streams carried in the 184 byte payload of a TS packet. There are four PSI tables carried in the MPEG-2 transport stream [1]:

- *program association table (PAT)*;
- *program map table (PMT)*;
- *conditional access table (CAT)*;
- *network information table (NIT)*.

These tables contain information needed for demultiplexing and presenting programs carried in the TS at the digital TV receiver. The PSI tables are carried in the MPEG-2 TS packet payloads using the MPEG-2 sections mechanism.

Many applications of MPEG-2 require storage and retrieval of ISO/IEC 13818 streams on various *digital storage media (DSM)*. A *digital storage media command and control (DSM-CC)* protocol is specified within ISO/IEC 13181 to facilitate the control of such media [1]. The applications of the DSM-CC protocol, including the DSM-CC sections mechanism and DSM-CC data and object carousels, are described in [2]. These applications enable the conveyance of private data in MPEG-2 transport streams.

This basic service provision defined by the ISO/IEC 13181 is augmented in the respective digital TV standards through the introduction of MPEG-2 TS transmission technologies for various physical transmission media, including satellite, cable and terrestrial systems and high-level mechanisms for broadcasting application data in the TS. The standards also define how interactive services are realized in the respective platforms including software APIs for end-user terminals and standards for the utilization of interactive media. The APIs for end-user terminals define a platform for the deployment of value-added services, with support for accessing resources from the digital

TV broadcasts and through a feedback network. The Sun Microsystems JavaTV API serves as a fundamental building block for the end-user terminal software APIs. As MPEG-2 is the basic technology for unidirectional digital TV broadcasts, an "all-IP" approach is the basic networking standard for interactivity-enabled services.

2.1.2 DVB

The European *digital video broadcasting (DVB)* consortium has developed a set of standards for MPEG-2 based digital TV establishing a platform currently being adopted in most of Europe, Asia and Australia. Basically, DVB has defined a system reference architecture for digital TV. An extension of the architecture is illustrated in Fig. 2.2.

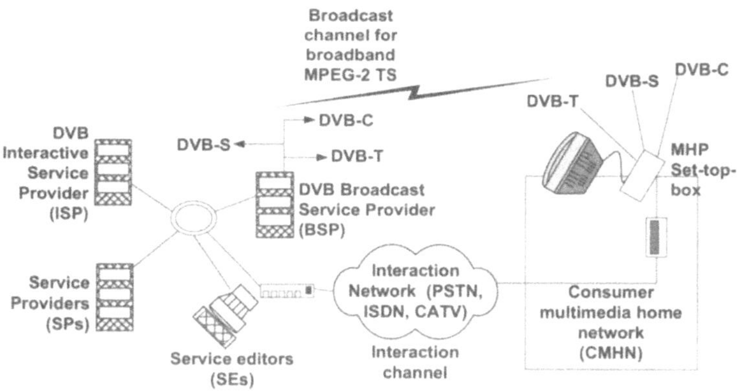

Fig. 2.2. An extended DVB system architecture (adapted from [132])

This extended DVB architecture has the following central components: a DVB broadcast service provider (BSP), a DVB *interactive service provider (ISP)*, *service providers (SPs)*, *service editors (SEs)*, a *broadcast channel*, a *feedback channel* and the *consumer multimedia home network (CMHN)*.

The BSP delivers broadband MPEG-2 transport streams over satellite, cable or terrestrial physical broadcast media to the consumer. The streams include standard MPEG-2 audio-visual content, *DVB service information (DVB-SI)* extending MPEG-2 PSI and annotating the stream contents, and value-added applications and data. The latter are transported over the TS with DVB data broadcasting. DVB enables the play-out of different profiles for *high definition television (HDTV)* or *standard definition television (SDTV)* at frame rates of 30 Hz or 25 Hz. PAL- or SECAM-compliant deployment requires the latter, which is the MPEG-2 *main profile at main level (MPML)*.

MPEG-2's bit-rates range typically from 4 Mbit/sec up to 9 Mbit/sec (CCIR-601 studio quality), at aspect ratios of 4:3, 16:9 and 2.21:1, and at luminance resolutions of 720x576, 544x576, 480x576, 352x576 or 352x288.

Service providers (SPs) create content and act as feedback network partners. A SP might be a travel agency, subcontracting electronic ticket selling to a BSP. The BSP would multiplex advertisements into the broadcast stream, upon which the consumer could buy electronic tickets from the travel agency directly over the feedback network.

At the consumer side the consumer multimedia home-network (CMHN) interconnects multimedia equipment in a home setting. The consumer accesses the TS services with a Multimedia Home Platform (MHP)-compliant set-top box, a standard part of the CMHN. MHP is an extension of the DVB digital TV standards for consumer set-top boxes. It basically defines a Java API for value-added services in a digital TV set-top box. For value-added applications implementing non local interactivity, DVB interactive services, a feedback channel is available for two-way information exchange. The feedback channel is provided by the ISP through various wired and wireless networks. IP-based protocols are typically used in the feedback channel.

Service editors (SEs) are responsible for creating the overall services, thus implementing the applications delivered by the broadcaster. The service editor creates advertisement materials and initializes movie production among other things.

The DVB system reference architecture (including the BSP, ISP, the consumer STB and the broadcast and feedback channels) relies on the following key standards:

- MPEG-2 TS transmission technology standards for different physical broadcast media;
- DVB-SI service information standards augmenting the standard MPEG-2 PSI information;
- DVB data broadcasting standards based on MPEG-2 and the DSM-CC protocol;
- DVB interactive service standards including DVB feedback channel standards for network independent and network dependent protocols;
- Miscellaneous other standards.

DVB has defined standards for the transmission of MPEG-2 transport streams in three physical broadcast media: satellite (DVB-S), cable (DVB-C) and terrestrial (DVB-T). The standards share many common technical elements enabling a high integration level for digital TV broadcasting and receiver equipment. However, due to the dissimilarities of the physical media, different signal modulation techniques are used resulting in different transmission capacities. DVB-S for satellite systems is the oldest and most widely used of the standards utilizing QPSK (Quadrature Phase-Shift Keying) modulation. DVB-C for cable systems is similar to DVB-S but uses QAM (Quadrature Amplitude Modulation) modulation. DVB-T for terrestrial digital TV utilizes yet another modulation technique known as COFDM (Coded Orthogonal Frequency Division Multiplexing).

DVB-SI [5] augments the standard MPEG-2 PSI with additional service information tables. These tables enable the identification of audio-visual and value-added services and events in a DVB streams. In contrast to the MPEG-2 PSI tables, DVB-SI tables can contain information about services in multiplexes (or even networks) other than those in which they are contained. As for PSI tables, the DVB-SI tables are segmented to sections before insertion into TS packets of the multiplex.

DVB data broadcasting [3] is an extension of MPEG-2 based DVB transmission standards. This extension provides ways to transport data other than standard MPEG-2 A/V data in DVB broadcast systems. Examples of data broadcasting include the download of software or the delivery of Internet services over a broadcast link. Four data broadcasting profiles are described within the standard based on the corresponding application areas with varying requirements for data transport.

- *data piping:* a simple and asynchronous mechanism for the end-to-end delivery of data in DVB broadcasts. Data are carried directly in the payloads of the MPEG-2 TS packets.
- *data streaming:* a mechanism for streaming-oriented end-to-end delivery of data in DVB broadcasts. Data are carried in MPEG-2 PES packets. Synchronization is also possible.
- *multiprotocol encapsulation:* a mechanism for the transmission of datagrams of communication protocols (e.g. TCP/IP) via DVB broadcasts. Datagrams are encapsulated in DSM-CC sections compliant with the MPEG-2 private sections format.
- *data carousels:* a mechanism for the periodic transmission of data modules in DVB broadcasts. The data carousel mechanism enables the update, removal and addition of modules to the data carousel. The DVB data carousel mechanism is based on the DSM-CC data carousel.
- *object carousels:* a mechanism for the periodic broadcasting of DSM-CC user–user objects in DVB broadcasts. The mechanism is based on the DSM-CC data carousel and DSM-CC object carousel mechanisms defined in MPEG-2 DSM-CC.

DVB specifies a number of network independent protocols for DVB interactive services and defines guidelines for their implementation and usage in [8] and [67], respectively. The DVB feedback channel allows the upstream flow of data from the users to the interactive service provider. Network dependent protocols for various physical feedback channel media are defined separately, e.g. for PSTN/ISDN in [7] and for CATV in [6]. The network independent protocols accessible in the feedback channel are either IP-based, MPEG-2 DSM-CC based or service specific. Network independent protocols in a broadcast channel for interactive services rely on the DVB data broadcasting standards.

In addition to the key standards introduced earlier, DVB digital TV standardization also covers DVB subtitling [66] and provision of standard ITU teletext services in DVB bit streams [65].

2.1.3 MHP

The *multimedia home platform (MHP)* adds a technical solution for the consumer receiver or STB enabling the reception and presentation of applications in an open and vendor, author and broadcaster neutral framework. Applications from various service providers will be interoperable with different MHP implementations in a horizontal market, where applications, networks and MHP terminals can be made available by independent providers [68]. In practice, MHP defines a Java-based STB API for value-added services in DVB broadcast systems. It relies on the DVB digital TV system reference model with standardized broadcast and feedback channels.

MHP covers three profiles: *enhanced broadcasting*, *interactive broadcasting* and *Internet access*. Enhanced broadcasting combines digital broadcast of audio/video services with downloaded applications that may use local interactivity. A feedback channel is not required in this profile. Interactive broadcasting enables a range of interactive services associated with or independent ofbroadcast services. This application area requires a feedback channel. Internet access is intended for the provision of Internet services. It also includes links between Internet services and broadcast services [68].

MHP Receiver

A layer model for an MHP receiver or set-top box, combining software and hardware components, is illustrated in Fig. 2.3.

The blocks on the right-hand side of the figure describe access to and use of basic digital TV A/V broadcast services (i.e. viewing different TV channels) of the MHP receiver. An MHP receiver includes as a standard feature a navigator application for selecting and tuning digital TV A/V services which may include interactive value-added services.

The layer model for value-added MHP applications is shown in the blocks on the left-hand side of the figure. Interactive application code and data are typically transported to the receiver through the DVB data broadcast DSM-CC object carousel mechanism over the broadcast MPEG-2 TS. The application code and data can also be uploaded over the interactive channel. An MHP receiver can support two types of interactive applications controlled by its application manager:

- Procedural DVB-J applications;
- Declarative DVB-HTML applications.

DVB-J is an adaptation of the Sun Microsystems' Java programming language and J2SE API to MHP receivers. DVB-J defines a DVB-J application model and a DVB-J platform.

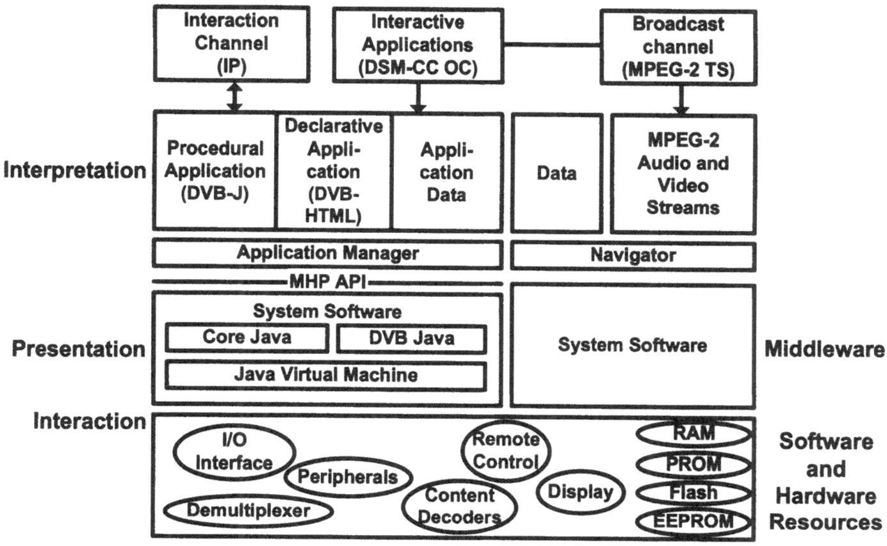

Fig. 2.3. MHP-compliant receiver architecture

The DVB-J application model defines a lifecycle and signaling model for DVB-J applications. The application model is based on the JavaTV API. The lifecycle control of a DVB-J application is based either on user interaction or on MHP application signaling carried in the broadcast *application information table (AIT)*. AIT is an extension of the DVB-SI to provide support for interactive value-added applications.

A DVB-J application is executed in the Java virtual machine, which is a standard part of the system software of an MHP receiver. The DVB-J platform consists of multiple APIs some of which are part of the standard J2SE API (Core Java) and which are specific to MHP (DVB Java).

The Core Java APIs include a subset of standard Java AWT graphical user interface classes and interfaces and other packages from J2SE such as networking and standard I/O services. Additional TV-oriented GUI components are depicted in the HAVi and JavaTV APIs. Components for playing streamed media are taken from the Java Media Framework API. Other classes and interfaces from the DAVIC and JavaTV APIs provide access to the DVB data broadcasting resources (e.g. DSM-CC object carousel file systems) and DVB-SI information. The DAVIC API also enables the tuning of digital TV services from DVB-J applications. The platform also includes other APIs for such matters as *conditional access (CA)* control.

The declarative DVB-HTML application is basically a set of DVB-HTML compliant HTML documents and content objects. DVB-HTML applications can be considered an adaptation of the World Wide Web to digital TV. MHP defines an application, lifecycle, and signaling model for DVB-HTML applications similar to those of DVB-J applications.[1]

The system software for both standard digital TV A/V services and interactive value-added applications relies on the hardware and software resources available in an MHP receiver. These resources include the physical broadcast and feedback channels, other I/O interfaces and local hardware for data storage, processing and display. Some of these components are illustrated in Fig. 2.3. Different MHP profiles and options enable the delivery to hardware platforms ranging from 30 MHz, 1–2 MB RAM and 1–2 MB Flash/ROM up to high-end devices [140].

2.1.4 Emerging DVB Standardization Efforts

As digital convergence breaks boundaries between the various digital media platforms, the DVB consortium is currently pursuing ways to bring the digital TV experience to a wider range of consumer devices.

May 2001 was the starting point for envisioning the future of DVB and its related standards. A strategy was developed to adapt DVB standards for the challenges of this century. The initiative is known as DVB 2.0 and its key points are: *further development of existing DVB standards*, *digital convergence*, *mapping of migration paths* and *globalization of DVB* [133].

Two specific action points related to DVB 2.0 are the *Globally Executable MHP (GEM)* specification and DVB-H.

GEM is a specification that harmonizes existing interactive digital TV initiatives under the umbrella of MHP. It seeks to make MHP a global and universal standard for interactive digital TV [72].

The DVB-H ("H" for hand held) standard seeks to provide access to broadband digital TV services from various mobile platforms. Digital TV on mobile platforms provides a possibility for broadcasters to reach their customers time and location independently while it enables mobile operators to implement data broadcast services in mobile service provision. For consumers the availability of all modalities of digital communication in a hand held device is a big step towards digital convergence.

2.1.5 ATSC-DASE and Open Cable

The *Advanced Television Systems Committee (ATSC)* is the digital TV standard of choice for North America and South Korea. A US organization, ATSC defines standards for terrestrial and cable digital TV broadcasting. Similarly to DVB, MPEG-2 is utilized for video encoding. However, non-MPEG encoding is used for digital audio. Also, the signal modulation techniques used differ from those of DVB. As in DVB, ATSC supports such methodologies as

[1] It is important to note, that in current MHP-compliant consumer device implementations, the whole DVB-HTML application environment is not supported.

data broadcasting. DASE or DTV Application Software Environment is similar to MHP in providing a platform for interactive value-added application on ATSC digital TV receivers. DASE is similar to MHP in its structure and implementation.

OpenCable is another US standard for digital cable TV. It includes both a hardware specification for physical cable digital TV receivers and a software specification. The software specification known as *OpenCable applications platform* or OCAP creates a common platform upon which interactive services can be deployed. The current versions of OCAP have adopted parts of the MHP specification. Furthermore, the harmonization of DASE and OCAP is an ongoing process. The OpenCable initiative is managed by Cable Television Laboratories, Inc. (CableLabs) which is a US-based nonprofit research and development consortium.

Version 1.0 of OCAP was published in December of 2001. OCAP is basically a software middleware layer providing access to operating system and hardware resources of an OpenCable digital TV host device (see Fig. 2.4). The primary business objective of OCAP is the introduction of a hardware and operating system agnostic interface for interactive digital TV application development for the US cable TV market. OCAP enables service and application portability between different OpenCable manufacturers' digital TV consumer devices and hopes to accelerate the development and deployment of interactive digital TV services in the US.

OCAP 1.0 includes a Java-based *execution engine (EE)* for interactive digital TV services and is similar to the MHP efforts of the European DVB consortium. In fact, OCAP 1.0 is built heavily on the DVB-MHP specifications with several OCAP specific extensions.

OCAP 2.0 released in April 2002 includes a *presentation engine (PE)* which includes an HMTL and ECMAScript engine for interactive hypertext-based applications. It also includes a bridge between the EE and PE for sharing programming environment resources between the engines.

2.1.6 ISDB-BML

Integrated services digital broadcasting (ISDB) has been adopted in Japan as the national digital TV standard. It includes specifications for satellite, terrestrial and cable digital TV systems. As for DVB, video and audio encoding is based on MPEG-2. However, the signal modulation techniques for the supported physical broadcast media differ again from those of DVB. The Japanese digital TV standard for interactive service provision is BML, which is basically an XML-based broadcast multimedia coding scheme and is similar in functionality to DVB-HTML.

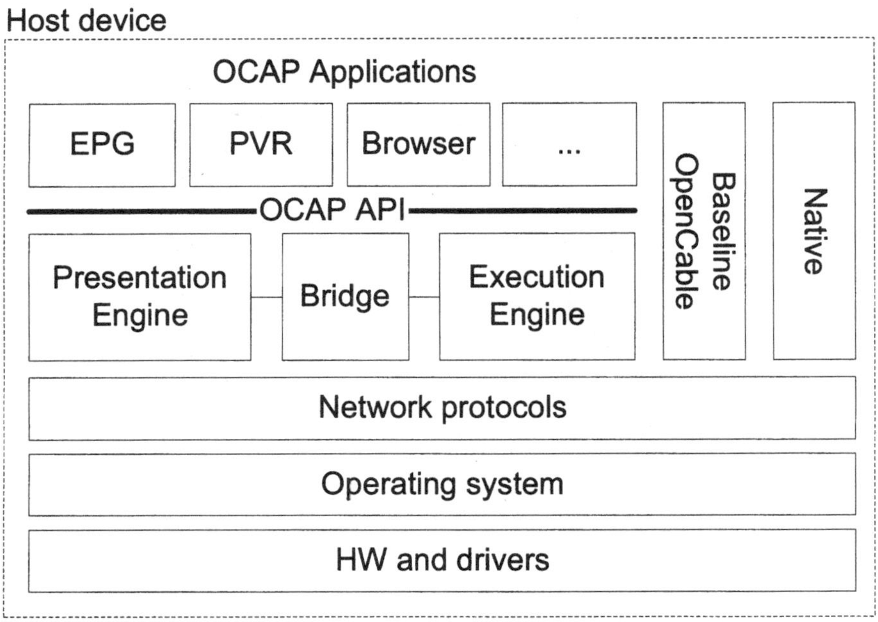

Fig. 2.4. The OCAP host device software architecture (adapted from a figure in "OpenCable Overview" presentation available at www.opencable.com)

2.1.7 Adoption of the Standards

Figure 2.5 illustrates the adoption of the various digital TV standards worldwide in the spring of 2003. As evident DVB-MHP is the most widely adopted standard.

2.2 Digital TV Asset Lifecycle

A common digital TV asset lifecycle is illustrated in Fig. 2.6 [2]. Its structure corresponds to the asset lifecycle of a generic networked multimedia asset lifecycle consisting of six parts:

- preproduction,
- production,
- postproduction,
- delivery,
- consumption and

[2] Andreas Mauthe and Oliver Morgan have used this figure in various presentations and publications.

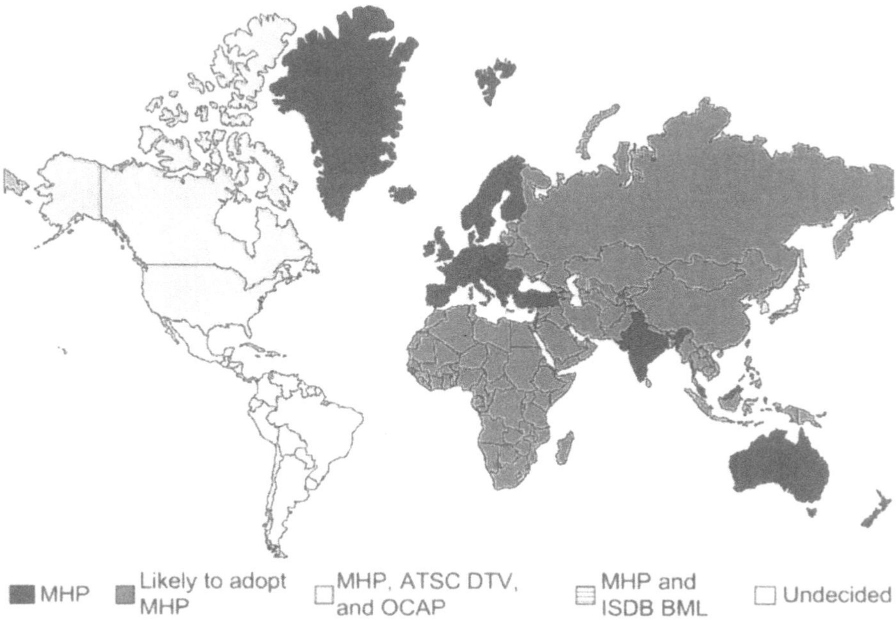

Fig. 2.5. Adoption of digital TV standards according to the DVB consortium [56]

- interaction and transaction.

We consider the asset lifecycle here in the context of a DVB-MHP digital TV where each part consists of actions needed to produce the generic service type: a digital A/V service combining value-added applications.

In the preproduction part the target service is defined and its product cycle is defined. The production part consists of the capture of the digital A/V content (i.e. the TV program) and development of value-added applications (i.e. MHP value-added applications) bound to it. In postproduction the captured digital A/V content is edited to its final form. In delivery, the finalized A/V content and the value-added applications are combined to a standard broadcast form, an MPEG-2 TS, for play-out in a DVB-MHP digital TV system using the facilities of the DVB broadcast service provider or BSP. The service is delivered to the consumers' MHP set-top boxes via a broadcast network. In the delivery stage the facilities of the ISP are also set up to provide full functionality for services implementing non local interactivity. Consumption is the stage where the consumer uses the developed service in different interaction modalities depending on the service type.

Service provision can be roughly divided into three basic business transaction scenarios within the asset lifecycle:

- *business-to-business (B2B)*,
- *service provider-to-service provider (S2S)*, and

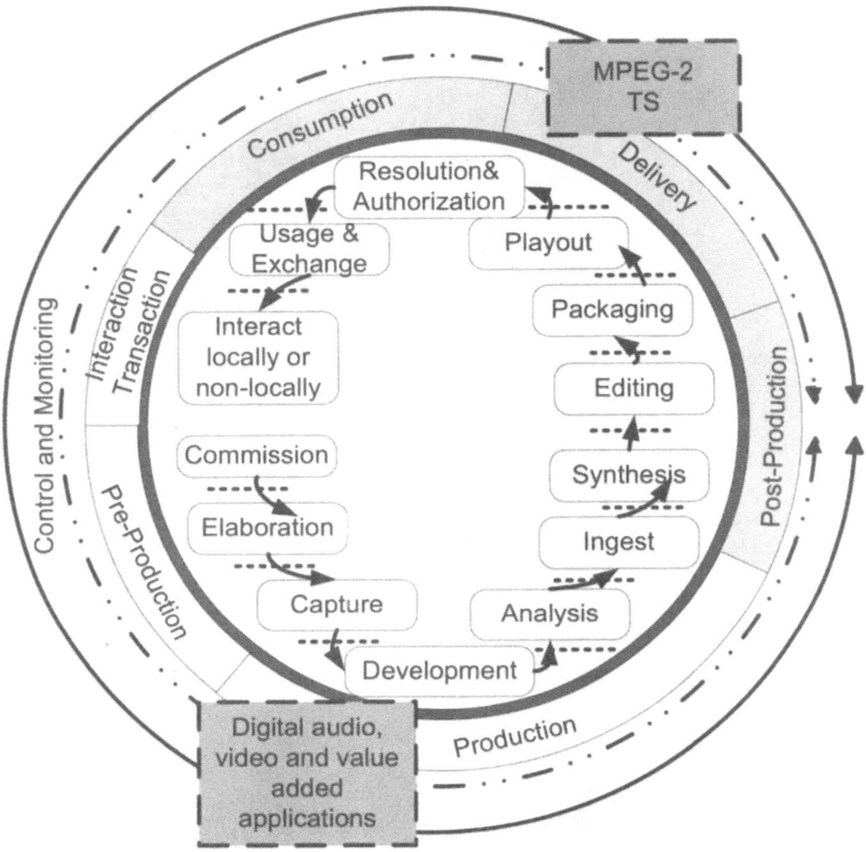

Fig. 2.6. Digital TV asset lifecycle (adopted from Andreas Mauthe and Oliver Morgan)

- *Business-to-consumer (B2C).*

In business-to-business (B2B) service provision the transaction is between two or more commercial entities. The entities have established a media service provision model within a digital TV system with the digital TV service providers realizing the technological solutions for the model. An example of B2B service provision in digital TV is a scenario where a digital content producer provides content for value-added services to be included in standard broadcast services.

In service provider-to-service provider (S2S) provision the transaction is between two or more digital TV service providers (i.e. ISPs and BSPs). An example of such a scenario is an agreement between a broadcaster and an interactive service provider to establish standard feedback channel functionality for digital TV services of the broadcaster.

The business-to-consumer (B2C) business transition is the scenario directly visible to the end-user. Here, a commercial entity and a consumer agree on a service to be provided to the consumer through digital television. Examples of such service include shopping, banking and health care services.

2.3 Examples of Digital TV Value-Added Services

Apart from the standard MPEG-2 audio-visual TV programming services, DVB-MHP compliant digital TV enables a range of value-added services. These services can fully utilize the rich content provision and interaction capabilities of the platform enabling truly innovative networked multimedia services.

2.3.1 Electronic Program Guide (EPG)

EPG or electronic program guide is the basic digital TV service including information about available TV programming. In conventional analog television EPG information is carried within the teletext services. In digital TV EPGs are directly based on the service information (SI) carried in the broadcast stream describing the available TV services. EPGs typically include service tuning functionality and simple content filtering based on e.g. program genre (see Fig. 2.7).

2.3.2 Information Portal

Conventional teletext services providing access to program information, news and other assorted simple text-based services have been the most popular value-added service in conventional analog television. Digital television helps to realize the next evolutionary step for teletext services with the introduction of information portal services also known as super teletext services. Realized as value-added applications information portal services provide significant improvements compared to conventional analog TV teletext. These include the availability of high-resolution graphics, improved formatting capabilities and a larger bandwidth available for carrying teletext data. In short, these services bring basic hypermedia capability to television. Typical contents of the information portal services include news (see Fig. 2.8), weather information and other simple hypermedia services.

2.3.3 Pay-per-View (PpV)

In PpV services the consumer purchases the rights to a one-time view of a particular piece of digital TV A/V content (e.g. a sports broadcast). Use of

Fig. 2.7. Screenshot of a typical EPG (© Ortikon Interactive Ltd.)

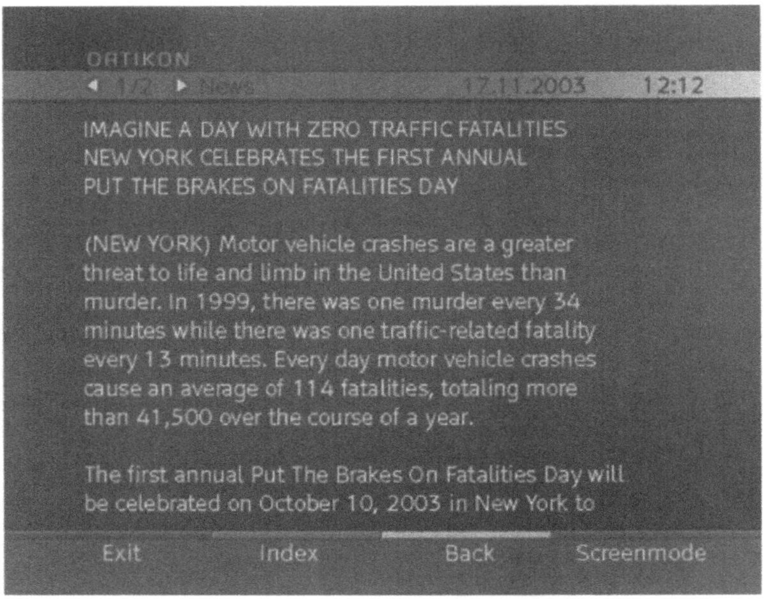

Fig. 2.8. Screenshot of a digital TV news service (© Ortikon Interactive Ltd.)

Pay-per-View within a digital TV system necessitates the use of a conditional access (CA) system for handling viewing authorization.

2.3.4 Video-on-Demand (VoD)

VoD enables consumers to order and view a particular piece of A/V content (e.g. a motion picture) with instant availability. VoD service provision requires the use of a high-bandwidth channel to transfer the content to the consumer receiver.

2.3.5 Education

Educational digital TV services range from traditional educational TV programming to interactive value-added services operating either in an independent context or in combination with an A/V service. Scenarios such as enhanced distance learning and education can be easily established within digital TV. With the use of a feedback channel a direct link between the tutored and the educator can be established.

2.3.6 Shopping

Digital TV provides an excellent medium to realize online shopping services where A/V content about products is combined with value-added services enabling their direct purchase. Basically, digital TV shopping enables, with a feedback channel available, full integration of TV shopping to the TV viewing experience. See Fig. 2.9 and Fig. 2.10.

2.3.7 Games

Simple computer games are easily implemented as digital TV value-added services (see Fig. 2.11 for an example of game service in digital TV).

2.3.8 Standard Internet Services

Digital TV value-added services can provide an interface to such standard Internet services as e-mail and limited Web browsing. See Fig. 2.13.

2.3.9 Communication

Services integrating current mobile messaging technologies such as text-based SMS and MMS multimedia messages to the digital TV platform can be realized. The consumer can send and receive these messages with a digital TV receiver. Both private two-way messaging and interactive message forums in a community setting are possible. See Fig. 2.12.

Fig. 2.9. Shopping in digital TV with shopping carts (© Ortikon Interactive Ltd.)

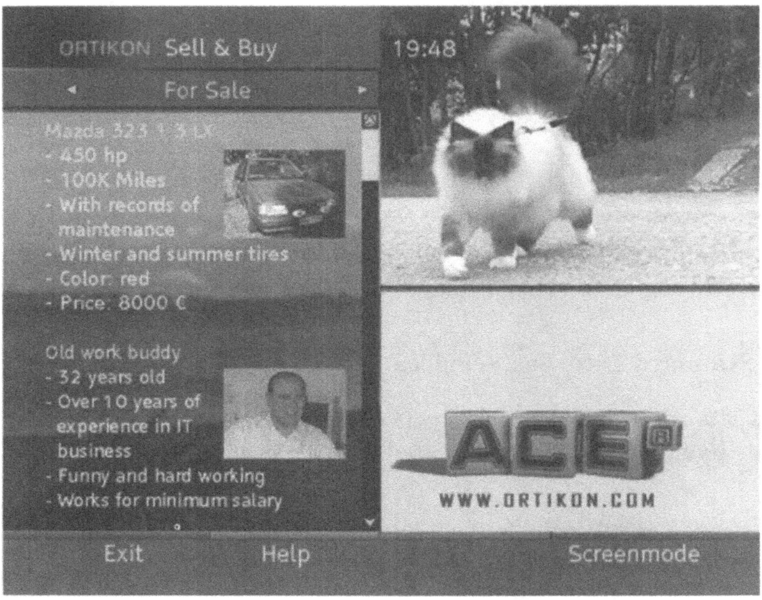

Fig. 2.10. Shopping in digital TV may involve both professional eShops and services for consumers to sell private goods (© Ortikon Interactive Ltd.)

Fig. 2.11. Screenshot of a Tetris-like game in digital TV (© Ortikon Interactive Ltd.)

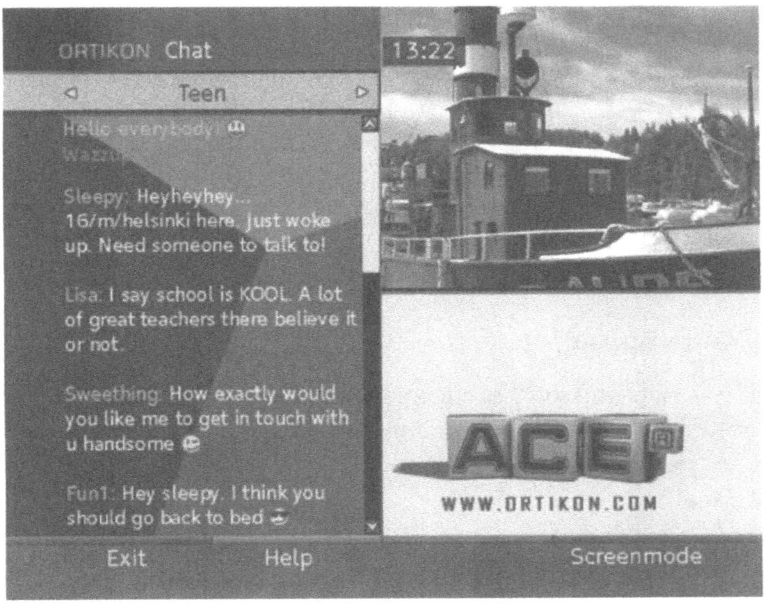

Fig. 2.12. A typical chatting service in digital TV (© Ortikon Interactive Ltd.)

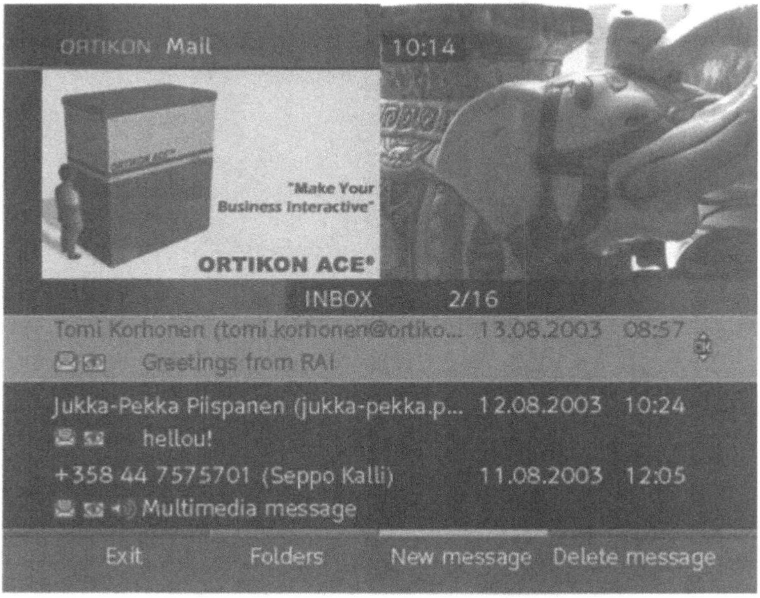

Fig. 2.13. Screenshot of an e-mail service in digital TV (© Ortikon Interactive Ltd.)

2.3.10 Community Services

Community services help create virtual communities in digital TV with the introduction of value-added services built for specific consumer groups with shared interests. They typically combine information portal and messaging services providing an avenue for specific audiences to observe value-added multimedia content relevant to their interests and to communicate with similarly oriented peers. See Fig. 2.14 for an example of a community service in digital TV.[3]

2.3.11 Government

Various governmental services currently available on the Internet can be similarly realized in digital television. These include online tax and voting services.

2.3.12 Health

In health care digital TV enables numerous innovative service schemes ranging from traditional health-related TV programs enhanced with interactive value-

[3] The concept of Sooda has been developed by Zento Interactive Ltd., Tampere, Finland. We would like to thank for providing the related materials and screenshots.

Fig. 2.14. Sooda is an online community service for young people in the age group 12 to 19 (concept developed by Zento Interactive Ltd. of Tampere, Finland, www.sooda.com, © Zento Interactive Ltd.)

added services to the use of a digital TV receiver as a terminal for telecare applications (see Fig. 2.15). Typical telecare applications include the exchange of information related to follow-up of various health conditions through digital television utilizing the feedback channel.[4]

2.3.13 Finance and Banking

Simple account management and brokering services are easy to implement within digital TV. The services can provide access to bank account data and the possibility for making various financial transactions similar to Internet banking.

[4] The concept of health care TV has been developed by Terivan Ltd., Tampere, Finland (`http://www.terivan.com`). We would like to thank them for providing the related materials and screenshots.

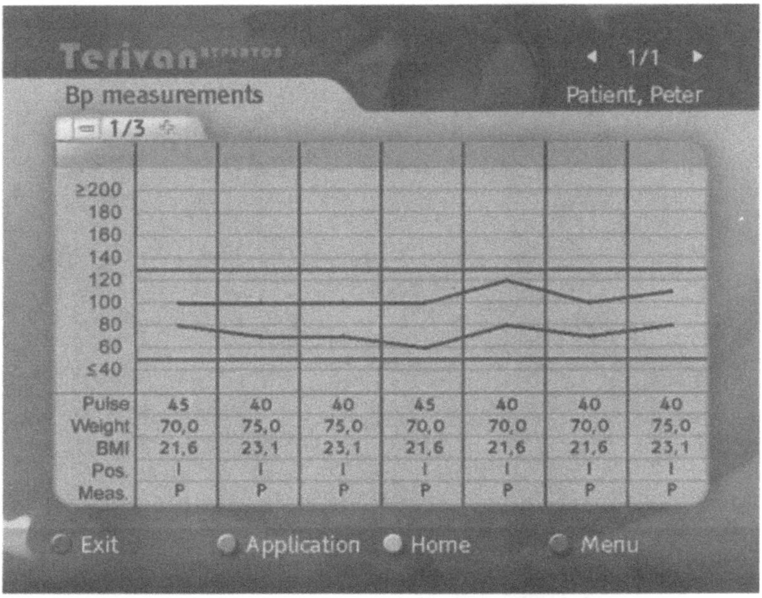

Fig. 2.15. Blood pressure monitoring in digital TV (concept developed by Terivan Ltd. of Tampere, Finland, http://www.terivan.com, © Terivan Ltd.)

3

Metadata Fundamentals and Concepts

Consumers learned interactivity on the Internet. Browsing through the Web is enabled by metadata, HTML, which structures and links the content of the Internet. The introduction of metadata in digital TV might lead to a new "killer application", as stated by Tom Worthington from the Australian National University [173]: "Existing digital broadcasting has a failed business model [as] it ignores what the audience wants most: control. Metadata has provided a hidden 'killer application' on the Internet and could be used to transform digital broadcasting into a viable service."

The reasons for metadata being a killer application in broadcasting are simple:

- interoperability across platforms within broadcasting value-chain partners;
- open and standardized interfaces for semi-automated multimedia asset sharing;
- harmonized and standardized exchange of data across the value-chain;
- rigidly annotated metadata guarantees high performance (e.g. fast DVB-SI parsing);
- higher granularity and higher level of detail with more complex metadata (e.g. making content narrative with metadata);
- scaleable temporal and spatial metadata resolution for content representation;
- bringing context into content representation is simplified with unified metadata models;
- presentational models for easier interactivity;
- easy and rapid adaptation, transcoding, transforming and encapsulation of metadata and content;
- metadata serves as a virtual nexus for the creation, delivery and exchange of multimedia assets throughout a virtual service space.

Within the scope of this chapter we present different types of metadata standards as typically used in broadcasting. We fill the gap between the theoretical foundations of metadata coming from the field of theoretical computer science and their practical application within MPEG-7 or MPEG-21. Both MPEG-7 and MPEG-21 are introduced and described at a later stage. Of special interest are the W3C metadata standards as the basis for the development of other more advanced data structures.

In short, this chapter is based on different theories laying the foundation for metadata-based services:

- theoretical aspects of metadata coming from the field of computational theory or formal language theory;
- introduction to more general multimedia metadata languages such as MPEG-7 or MPEG-21;
- revisal of metadata standards typically used in broadcasting such as GXF, MXF, AAF, SMPTE, TV-Anytime etc.;
- convergence of different metadata types with the help of a general metadata map.

Current digital TV standards used widely in broadcasting focus purely on the definition of rigid metadata. They are easier to handle, faster to parse, easier to create, are streamable and are less complicated than granular metadata. However, more complex services than simple DVB-SI metadata-based EPGs are emerging. These and more advanced service types — e.g. narrative digital TV — would be difficult to realize. Therefore there is a direct need for more flexible metadata structures than rigid metadata for more complex use scenarios. This chapter focuses on the description of the fundamental granular metadata standards available for broadcast use. At the end of the chapter we map out the positions of these standards within a metadata enabled digital TV value-chain and their dependencies.

3.1 Digital TV Metadata Lifecycle

Figure 3.1[1] illustrates the different life-cycle phases during the production of a digital television show and assigns typically used metadata definitions to them. *Preproduction* focuses on planning, conceptualization and on continuous refinement of project ideas. Its outcome is more related to concepts, decision making and to the definition of basic requirements. Principal decisions and requirement documents guide further phases.

During *production*, pieces of multimedia assets in any form are concretized and realized. This might be the shooting of a film, implementation of services, creation of sound effects, establishing creational patterns and libraries among other things. *Postproduction* compiles the actual show and combines different multimedia assets into one piece for distribution (*delivery*).

A TV station can add its informational services to a news broadcast and organize the play-out to the consumers. Based on access rights and pay-TV

[1] It is important to note, that in current MHP-compliant consumer device implementations, the whole DVB-HTML application environment is not supported.

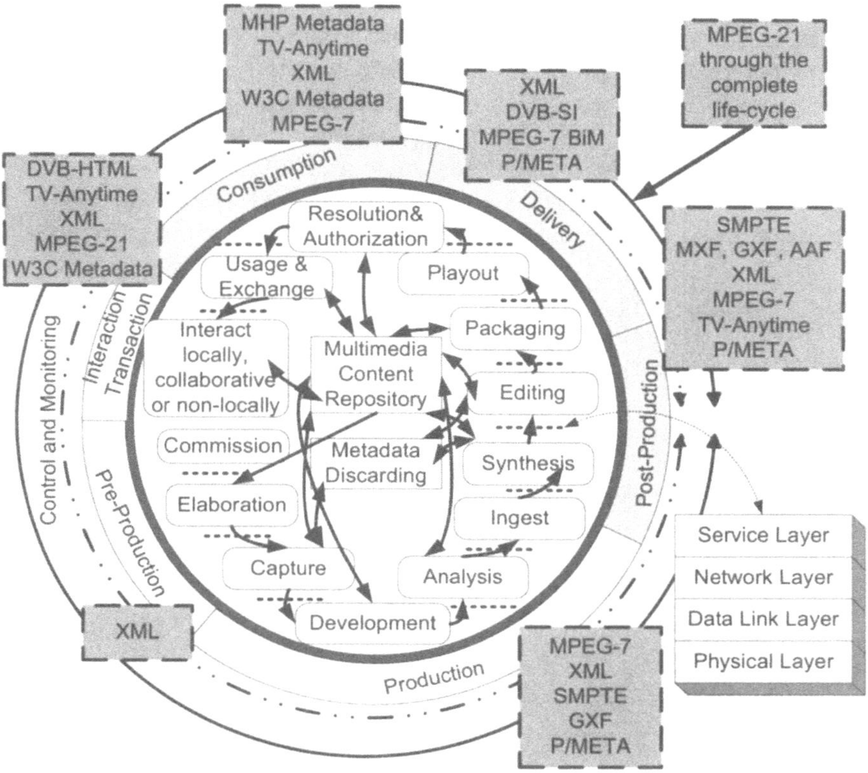

Fig. 3.1. Modified and extended version of Oliver Morgan's principal metadata life-cycle model (adopted from Andreas Mauthe and Oliver Morgan)

options the consumer can record the content with a VCR or enjoy the synchronized informational services. News broadcasts might be "spiced up" with this type of informative service. A feedback channel equips the consumer with interaction facilities during the *consumption* phase.

3.2 Theoretical Foundations of Metadata

"We are on the verge of a metadata revolution. Get your data models clean and prepare for an interesting ride...," stated Tim Berners-Lee in 1999 as quoted in [80]. However, looking at metadata from a narrow perspective makes it difficult to grasp its full picture. Therefore we explain metadata from different viewpoints. Within the scope of this chapter we approach the problem from the viewpoint of the theoretical computer science community. Formal languages, compiler construction and complexity theory lay the basics for representing and dealing with metadata. Further details can be found in [18, 12, 98].

A more practical approach comes from the type definition community, putting theory into specific application areas, such as image processing. The service developer community builds applications adequate for end-users and consumers, such as search engines or semi-automatic annotation tools. Furthermore the data warehousing community deals with storage, management, data models and filtering methods to obtain knowledge of stored records.

3.2.1 Metadata Tier Model

A model for categorizing different approaches and viewpoints unifying the theory of metadata is presented within this chapter and shown in Fig. 3.2. Each higher tier encapsulates knowledge and characteristics based on lower tiers. A lower tier is capable of describing higher layers. Each tier is self-containing, has its own atomic unit and is self-describing. Each model set operates on its own predefined rules, atomic units and characteristics.

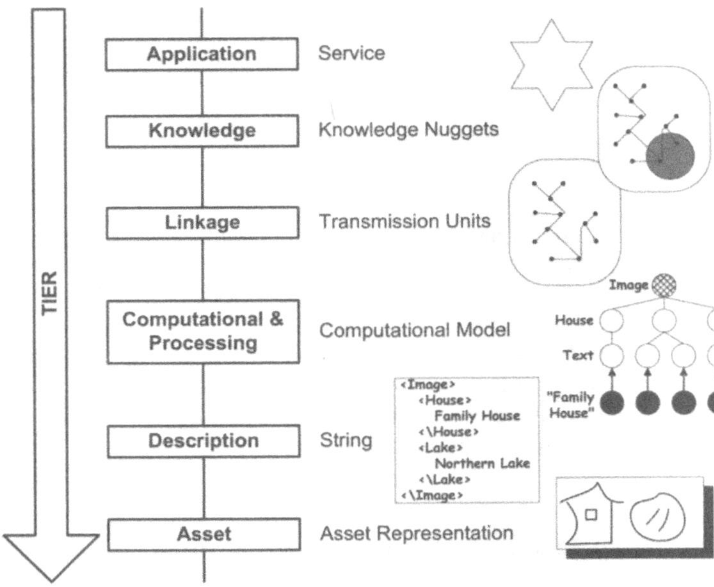

Fig. 3.2. Metadata tier model

- The *content tier* defines digital representations of multimedia content. In a wider sense it can also be called an *asset tier* by involving metadata assets in addition to content assets. In this case metadata assets are seen as entities to be dealt with as a whole rather than to act as a description of multimedia content (e.g. metadata package describing the content of an image). Its atomic units are complete multimedia entities or sub-components. The underlying model depends on the nature of the multimedia asset (e.g. image compressed in a specific format).

- The *description tier* is built on top of the asset or content tier and should build an abstraction with any granularity level of it. A very simple example is a Web page as a compilation of images brought into structure by HTML code. On this level the definition of languages of how to create rules, characteristics, sequences and properties of strings is dominant. On this layer the application developer community creates application-oriented data structures (e.g. MPEG and its MPEG-7 and MPEG-21 standards). The specific group around the theoretical computer scientists lays the theoretical foundations for metadata languages (e.g. W3C and XML).
- The *processing tier* or *computational tier* describes how to generate, validate and enumerate valid sequences of strings conforming to the definition. Top-down and bottom-up viewpoints exist. Theoretical computer science delivers the knowledge of how to apply metadata languages, parse data, create valid tree representations, etc. Mathematical models and algorithms deliver the relevant data to fill the leaves of created trees with data obtained from multimedia assets. The service development community parses, changes and applies the metadata trees within its applications.
- Linkage of metadata definitions and their sub-parts correlations are defined within the *linkage tier*. In a narrow sense the linkage tier describes how descriptions and their sub-parts relate to each other and in a wider sense with the transmission, synchronization and interlinking of metadata definitions. A simple example is the whole Web, being an interlinked network of single metadata documents in the form of Web pages. Search engines act as their interlink points.
- The meaning of data to specific users or user groups is given by models within the *knowledge tier*. This tier defines mathematical theory by giving semantic meaning and creating useful knowledge. Algorithms automatically crawling through the Web and presenting personalized Web pages based on user profiles are just one example.
- Finally, the *application tier* manifests intellectual property within computer programs, reveals the relevancy of metadata models and builds consumer-driven services around them.

3.2.2 Theory Behind the W3C Metadata Definition Family

The basis for the theory behind several metadata definitions comes from the theoretical computer science domain.

Each metadata language is based on a set of *strings* with certain characteristics. A *grammar* (G) provides a set of rules for constructing metadata trees based on strings. The overall set of valid instantiated trees based on a grammar is called a *language* (L). Each tree consists of a set of potential start symbols (S), which represent the root node of several parsing and creation processes. Terminal symbols (V_t) are tree leaves and non terminal symbols (V_n) intermediate tree nodes.

The *Chomsky Hierarchy* categorizes different classes of languages by defining phrase structure grammars and their properties. Noam Chomsky introduced the model in 1959.

Each phrase structure grammar is a four-tuple $G = (V_n, V_t, R, S)$, where V_n is an alphabet of non terminal symbols; V_t is a alphabet of terminal symbols; R a set of replacement rules in the form of $\alpha \rightarrow \beta$, where $\alpha \in V * V_n V*$, $\beta \in V*$; $S \in V*$ is a start symbol; and $V_n \cup V_t = V$.

Type	Language Type	Language Structure	Computational Model	More Power through Non determinism	Production Rules	Closure Properties	Characteristic Dependencies
0	recursively enumerable	**recursively enumerable grammar (REG)**	Turing machine	no	all	all	all
1	context sensitive	**context sensitive grammar (CSG)**	linear bounded automaton (Turing machine with a bounded tape)	yes	$\alpha X \gamma \rightarrow \alpha\, \beta$	$\cup, \cdot, *$	crossing
2	context free	**context free grammar (CFG)**	pushdown automaton	no	$u \rightarrow v\, w$ $u \rightarrow X$	$\cup, n\,\sigma\cap, \cdot, n\,\sigma-, *$	nested
3	regular	**regular grammar (RG)**	finite state automaton	no	$u \rightarrow X\, \iota$ $u \rightarrow X$	$\cup, \cap, \cdot, -, *$	no long-range

Fig. 3.3. Chomsky Hierarchy (as shown in [18])

The theory behind the W3C metadata definition family, especially XML schema [166] and the DTD [30] has been excellently researched by M. Murata and M. Takahashi. The definition part of the languages belongs to the description tier, their enumeration and automata for validation to the processing tier. They are based on *regular tree grammars and languages* [36]. DTDs are restricted to *local tree grammars and languages* [164] and XML schema correspond to *single type tree grammars and languages* [135].

In regular tree languages and tree automata [36] each node may have an infinite number of children nodes and the right-hand side of a replacement rule in the form of regular expressions over V_n of the Chomsky Hierarchy. Each replacement rule is in the form $A \rightarrow \alpha$, where $A \in V_n$ and α is a regular expression over non terminal and start symbols ($V_n \cup S$).

Definition 3.1 (regular tree grammar and languages). *The four tuple* $G = (V_t, V_n, S, R)$ *is a* regular tree grammar, *where* R *enumerates languages* *in the form* $A \to a\alpha$. $A \in V_n$ *is the left-hand side and the right-hand side is constructed by* $a \in V_t$, *in combination with a regular expression* α *over* V_n. *A regular tree language is enumerated by a regular tree grammar.*

Definition 3.2 (competing non terminal symbols). Non terminal symbols *(e.g.* X *and* Y*) are competing, if* $X \to a$ *and* $Y \to a$, *where* $a \in V_t$ *and* $X, Y \in V_n$.

Definition 3.3 (local tree grammar and languages). *A regular tree grammar with the restriction not to have competing non terminal symbols (*V_n*) is called a* local tree grammars. *A* local tree language *is enumerated by a local tree grammar.*

Example 3.4 (local tree language). A local tree grammar $G(L) = \{V_t, V_n, R, S\}$, $V_t = \{0, 1\}$, $V_n = \{A\}$, $S = \{S\}$ and $R = \{S \to 0A, A \to 100A, A \to \epsilon\}$.

Definition 3.5 (single type tree grammars and languages). *A* single type tree grammar *has non competing non terminal symbols (*V_n*) and non competing start symbols (*S*). A* single type tree grammar *is enumerated by a single type tree language.*

Definition 3.6 (attributed grammars and languages). *An* attributed grammar *extends other grammar types by the feature that each non terminal symbol can be annotated with attributes of different types (e.g.* $A_{i:int}$*) for syntax tree generation. An* attributed language *is generated by an attributed grammar.*

3.2.3 Practical Example

This example illustrates the purpose of the different layers of the metadata layer model and shows how XML, DTD and XML schema are applied. The source code for this example can be found in Tables 3.1 and 3.2.

Figure 3.4 shows an example image of two persons shaking hands. The XML schema defines the structure, how the image should be annotated. An instantiated XML file shows one possible metadata definition instance. Semantics add meaning, in our case the handshake of the two persons. The visualization of the instantiated metadata definition would belong to the application layer. Example 3.7 illustrates the related DTD as a local tree grammar.

Example 3.7 (DTD as local tree grammar). *#PCDATA* expresses in XML notation any arbitrary node content of a non terminal symbol, denoted by the terminal symbol *#pcdata*:[2]

[2] "*" denotes from 0 to ∞, "+" denotes 1 up to ∞ appearances.

Table 3.1. XML schema source file based on the example stated in this chapter (SimpleImage.xsd)

```
<?xml version="1.0" encoding="UTF-8"?>
<xs:schema xmlns:xs=http://www.w3.org/2001/XMLSchema
  <xs:element name="Image">
    <xs:annotation>
       <xs:documentation>Root element</xs:documentation>
    </xs:annotation>
    <xs:complexType>
      <xs:sequence>
        <xs:element ref="PersonDescriptor"
           minOccurs="0" maxOccurs="unbounded"/>
        <xs:element ref="ObjectDescriptor"
           minOccurs="0" maxOccurs="unbounded"/>
        <xs:element ref="TextualDescriptor"
           maxOccurs="unbounded"/>
      </xs:sequence>
      <xs:attribute name="ID" type="xs:string"
        use="required"/>
    </xs:complexType>
  </xs:element>
  <xs:element name="PersonDescriptor">
    <xs:complexType>
      <xs:sequence>
        <xs:element ref="TextualDescriptor"
          minOccurs="0" maxOccurs="unbounded"/>
        <xs:element ref="PositionDescriptor"/>
        </xs:sequence>
    </xs:complexType>
  </xs:element>
  <xs:element name="ObjectDescriptor">
    <xs:complexType>
      <xs:sequence>
        <xs:element ref="TextualDescriptor"
          minOccurs="0" maxOccurs="unbounded"/>
        <xs:element ref="PositionDescriptor"/>
     </xs:sequence>
    </xs:complexType>
  </xs:element>
  <xs:element name="TextualDescriptor" type="xs:string"/>
  <xs:complexType name="PositionDescriptionType">
    <xs:attribute name="x" type="xs:integer"/>
    <xs:attribute name="y" type="xs:integer"/>
  </xs:complexType>
  <xs:element name="PositionDescriptor"
     type="PositionDescriptionType"/>
</xs:schema>
```

Table 3.2. An instantiated XML file based on the example stated in this chapter (SimpleImage.xsd)

```
<?xml version="1.0" encoding="UTF-8"?>
<Image  xmlns:xsi=http://www.w3.org/2001/XMLSchema-instance
   xsi:noNamespaceSchemaLocation ="
   ...\SimpleImageSchema.xsd" ID="10">
  <PersonDescriptor>
    <TextualDescriptor>Anurag</TextualDescriptor>
    <PositionDescriptor x="25" y="25"/>
  </PersonDescriptor>
  <PersonDescriptor>
    <TextualDescriptor>Artur</TextualDescriptor>
    <PositionDescriptor x="75" y="75"/>
  </PersonDescriptor>
  <PersonDescriptor>
    <TextualDescriptor>Florina</TextualDescriptor>
    <PositionDescriptor x="75" y="25"/>
  </PersonDescriptor>
  <ObjectDescriptor>
    <TextualDescriptor>Coffe Cup</TextualDescriptor>
    <PositionDescriptor x="50" y="60"/>
  </ObjectDescriptor>
  <ObjectDescriptor>
    <TextualDescriptor>Table</TextualDescriptor>
    <PositionDescriptor x="50" y="50"/>
  </ObjectDescriptor>
  <ObjectDescriptor>
    <TextualDescriptor>Chair</TextualDescriptor>
    <PositionDescriptor x="25" y="50"/>
  </ObjectDescriptor>
  <TextualDescriptor>
   This image contains 6 objects
  </TextualDescriptor>
</Image>
```

$$\begin{aligned}
\Sigma &= \{Image, ObjectDescriptor, TextualDescription,\\
&\quad PositionDescription, PersonDescription, \#PCDATA\}\\
V_t &= \{\#pcdata\}\\
V_n &= \{Image, ObjectDescriptor, TextualDescriptor,\\
&\quad PositionDescriptor, PersonDescriptor\}\\
S &= \{Image\}\\
R &= \{Image \rightarrow
\end{aligned}$$

$$PersonDescriptor * ObjectDescriptor * TextualDescriptor+,$$
$$PersonDescriptor \rightarrow TextualDescriptor * PositionDescriptor,$$
$$PositionDescirptor_{x:int,y:int} \rightarrow \epsilon,$$
$$TextualDescriptor \rightarrow \#pcdata|\epsilon\}$$

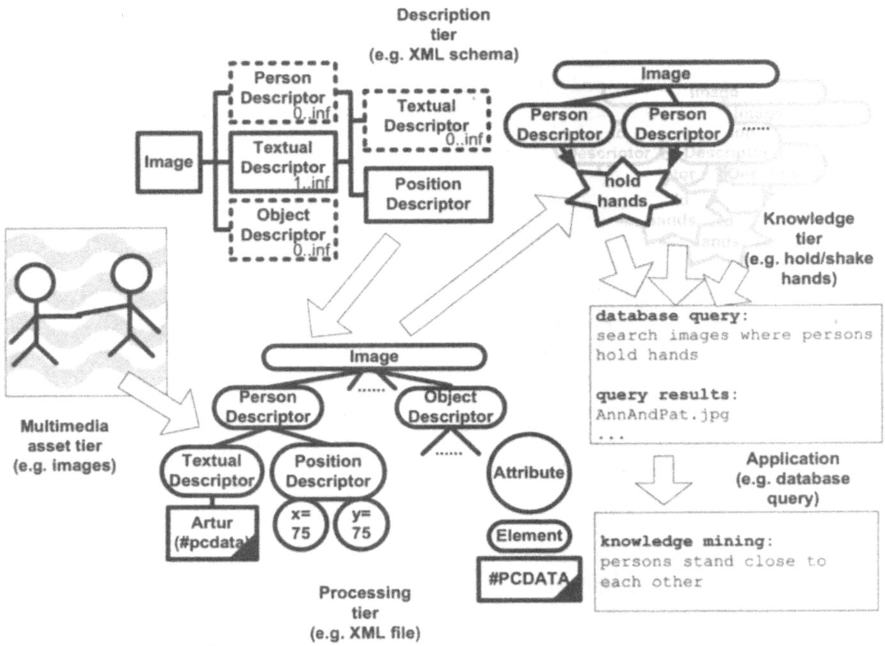

Fig. 3.4. Tier model: asset tier (image), description tier (schema), processing tier (XML file), linkage tier (reference trees), knowledge tier (handshake) and its application (visualization)

3.3 W3C Metadata Families

Tim Berners-Lee founded the W3C in 1994 to explore and shape the future development of the *World Wide Web (WWW)*. More than fifty specifications backed up by an international consortium consisting of over 450 member organizations guarantee interoperability. The efforts of W3C have resulted in commonly used specifications such as HTML [37], XML [39], SVG [41], etc.

High stability and the widespread adoption of W3C efforts in standardizing a metadata language has resulted in a growing common acceptance of *eXtensible Markup Language (XML)* by other standardization bodies. XML acts as a foundation for other metadata standards, such as MPEG-7, MPEG-21 or MXF explained within the scope of this chapter.

3.3.1 Overview of the W3C Metadata Families

A complete description of each metadata standard defined by the W3C is beyond the scope of this chapter. A complete list including reference software, standards and examples can be found on the W3C Web page (`www.w3c.org`). Table 3.3 gives a brief overview of the W3C definitions and their functionality.

Table 3.3. An overview of W3C metadata standards

Name	Description	Key-Features
XML	umbrella for XML schema	working groups for the development of the XML base platform
XML schema	metadata definition language	basic data types, document structure, etc.
XSLT	transformation of metadata	transcoding language for XML
DTD	metadata definition language	Similar to XML schema but more restricted
SVG	Scalable vector graphics, light-weight 2D vector graphics	light-weight 2D graphic representation of basic shapes, animations, transparency, images, etc.
SOAP	simple object access protocol, metadata exchange and encapsulation	metadata-based messaging, binary attachments, remote process invocation
XPath and XQuery	search, query and referencing XML documents	interlinking and referencing within XML documents
XLink	interlinking XML documents	creation of a network of links between documents (e.g. hyperlinks)
DOM	document object model, tree representation of XML files	dynamic software interface for document updates
CSS	cascading style sheets, style sheets for metadata visualization	definition of styles for the visual appearance of XML documents (e.g. fonts, color, etc.)
SMIL	synchronized multimedia integration language, multimedia presentation language	multimedia presentation features for presentation scheduling, timing, animations
HTML	Internet Web pages	visualization of Web content
RDF	Resource Description Framework	adding ontology for the integration of various XML-based exchange formats

The following is a classification of the W3C metadata definitions according to their underlying functionality and purpose:

- *Internet-type usage:* deployment of services in feedback channel equipped networks are typical in the Internet for push/pull services (e.g. HTML browser on a high-performance digital TV consumer device with access to the Internet);
- *broadcast channel services:* encapsulation of W3C services into a broadcast stream for push content only (e.g. instantiated XML files as configuration for services);
- *basic toolset for other metadata standards:* W3C definitions are an excellent source for developing new metadata standards for novel purposes (e.g. XML schema was selected as the basis for MPEG-7; and XHTML in DVB).

Applying W3C metadata standards in digital television mostly focuses on the deployment of services in feedback channel equipped consumer networks. Most of them belong to the consumption or interaction phase in the metadata life-cycle.

Multiple standards explore the different service types and application scenarios. Altogether they provide some common characteristics:

- comprehensive built-in and user-definable data types;
- unique identification via assignment of unambiguous namespaces;
- reusability and easy integration in XML-enabled architectures;
- event and tree-based metadata representation or metadata exchange.

3.3.2 XML

XML is a collection of standards based on *Standard Generalized Markup Language (SGML)* [9]. SGML was basically intended for use in digital publishing and manifests itself nowadays in many XML-based metadata definitions. The W3C foresees the basic functionalities:

- *metadata representation and transformation* of objects utilizing XSLT [38] for metadata transformation, XSL Formatting Objects (XSL/FO) and stylesheet representation;
- *metadata linkage and querying* includes the XML Linking Language (XLink) [50], XML Pointer Language (XPointer) [51] and the XML Query Language (XQuery) [44];
- *metadata definition language* is provided by XML schema defining document structure [168], data types [43] and use of metadata definition languages [42].

The next section concentrates on the descripton of XML schema as the basis for the creation of instantiated XML documents.

3.3.3 XML Schema

An XML schema defines the outlook, structure and data types defining how an instantiated XML file is structured. The purpose of this section is the clarification of constructs used to build XML schemas. Special focus is given to data types relevant for building an XML schema and in creating XML file instances.

Table 3.4. XML schema constructs

Construct	Subconstructs	Example
identifier	namespace	`xmlns:mpeg7=http://...`
	inner-document references	`<... id="element10".../>`
types	principal types	
	simple types	element
	complex types	element with children and attributes
	built-in types	float, double, integer, tokens, strings, etc.
structures	elements	`<element name="Image">...</element>`
	attributes	`<attribute name="ID" type="xs:string"/>`
	derived types	`<extension base="string">...</extension>`
	compounding constructs	`<group name="contacts">...</group>`

An XML schema consists of various constructs that can be roughly categorized (Table 3.4 shows an overview of schema constructs):

- *identifier* for unambiguous identification of whole XML schema documents and their subcomponents. Unique assignment of namespaces and inner-document identification of XML schema constructs via assignment of identifiers in the form of attributes belong to this set of constructs. A namespace is represented conforming to URI [23] and consists of a prefix and a string. The prefix identifies the overall schema and the string to document sub components. A target namespace identifies the target namespace of the newly created XML schema;
- *types* definitions can be divided into *principal types* and *built-in types.* Principal type definitions belong to *simple types* that do not allow child elements or attributes; *complex types* allowing child elements or attributes either with or without defining constraints on their values; and *built-in types* providing a basic set of tools for creating new type definitions or instantiated XML files (e.g. string, double, tokens);
- *structures* are constructs helping to extend, wrap or group existing document components in various ways. The most significant constructs are *elements* and *attributes.* Elements are assigned with a type and attributes associated with a data type are a part of elements. With both constructs it is possible to generate complex definitions together with other structural constructs. More complex structural components are *derived types* either extending or restricting existing type definitions and *compounding*

constructs for grouping elements or constructs (e.g. list type, union type, facets, group definitions).

Several components are utilized to build complete XML schemas and to instantiate them to XML files. Table 3.1 and Table 3.2 show an example of an XML schema and its instantiated counterpart.

3.4 MPEG-7 — Multimedia Content Description Interface

In 1988 the *Moving Picture Experts Group (MPEG)* was founded and started to define a series of standards covering multiple multimedia issues. First efforts resulted in digital audio/video compression standards, such as MPEG-1, MPEG-2 and MPEG-4. MPEG followed two other trails by defining MPEG-7 as a description standard for audio-visual content and MPEG-21 as an on-going packaging and content protection standard for multimedia assets.

MPEG-7 was first introduced in 1997. It defines a set of descriptions for audio-visual content in the form of *Multimedia Description Schemes (MDS)* for describing audio-visual content. The application level context for applying MPEG-7 in digital TV ranges from education, journalism, tourist information, cultural services, entertainment, shopping and audio-visual content production to film, video and radio archives. MPEG-7 standards and the excellent work called "Introduction to MPEG-7" [130] are the foundations of this chapter.

The standard has eight different parts, where the first five parts are normative and parts six to eight support:

- part 1 (*Systems*) standardizes binary transmission, synchronization and storage modes [116];
- the metadata language is defined in part 2 (*Description Definition Language*) [117];
- visual and audio description schemes are defined in part 3 (*Visual*) and part 4 (*Audio*) [118, 114];
- non-A/V content descriptors are described in part 5 (*Generic Entities and Multimedia Description Schemes (MDS)* [115];
- part 6 (*Reference Software*) [113] defines general software that supports the different standardized parts;
- part 7 (*Conformance Testing*) focuses on processes for testing MPEG-7 conformant hardware or software implementations; and
- part 8 (*Extraction and Use of MPEG-7 Descriptions)* defines the multimedia content description interface and the procedures for the use of MPEG-7 tools and the implementation of the reference software.

3.4.1 Overview

The growing amount of content requires a cost-efficient and format independent description language for annotating multimedia information. To achieve this goal MPEG-7 defines a minimal and format independent set of tools to maintain interoperability and flexibility. Definition of minimal standardized parts guarantees openness to competitive market forces of non normative elements of the MPEG-7 standard. Only the format of content descriptions is defined. The development of techniques for extraction, encoding and use is left to industry. This is also a challenge with the MPEG-7 type of metadata as automated extraction from metadata is often relatively difficult.

A wide area of data types based on XML is considered for describing content on high abstraction levels. MPEG-7 is based on the following basic principles and application areas:

- Annotation of multimedia content for search, retrieval and management of content;
- MPEG-7 can be characterized as a catalyst and type set source for newly emerging metadata standards;
- System implementation for textual representation and binary streaming of metadata.

The first one makes MPEG-7 applicable in its first intended use scenarios in multiple domains, such as education, broadcasting, biomedical applications, entertainment, multimedia content archives, etc. MPEG-7 as a catalyst and type set source allows new standards to be defined on top of MPEG-7 and to be refined to specific use-scenarios (e.g. TV-Anytime is based on MPEG-7 metadata definitions).

Hyperlinked TV is an example service, where a video stream is annotated with MPEG-7 descriptors. Hyperlinks are assigned to each frame by adding region descriptors. This example is described in further detail within the scope of this book. Figure 3.5 shows the principal MPEG-7 document for this application.

This section reviews MPEG-7 as a metadata definition language, presents an example, pinpoints broadcasting specific use-scenarios and is a guide through its rich data types.

3.4.2 MPEG-7 Metadata Definitions

In principle, MPEG-7 metadata definitions can be categorized as:

- *descriptor definition language (DDL)*;
- *multimedia description language (MDL)*;
- *visual and audio metadata definitions.*

The XML schema has been selected with slight extensions as the DDL. It builds the basics for MPEG-7 metadata definitions. The DDL describes a set

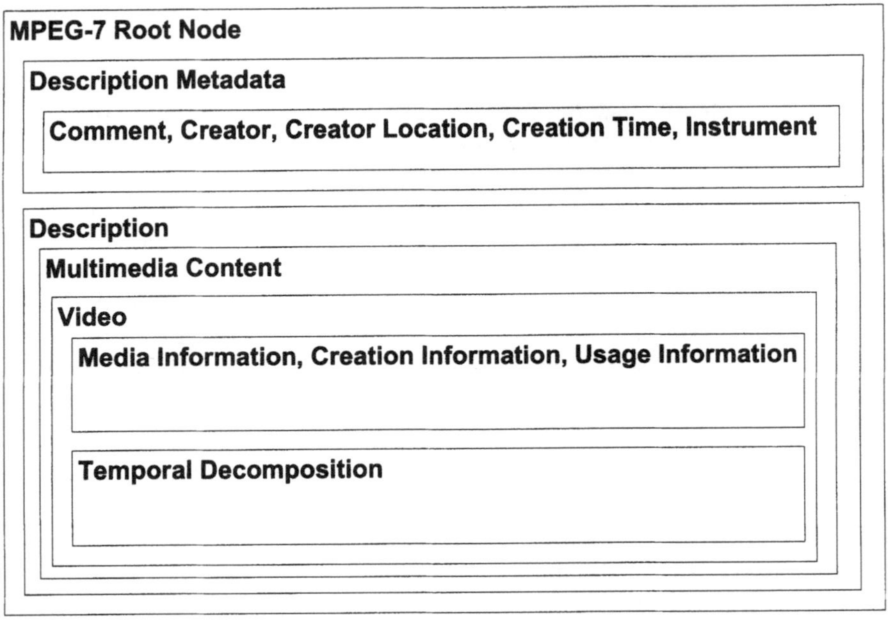

Fig. 3.5. Schematic description of the XML document for segmented TV

of rules for how to structure, extend, create and modify MPEG-7 metadata definitions in the form of *description schemes (DS)* and *descriptors (D)*.

Definition 3.8 (descriptor). *Syntactic and semantic representation of a feature of a multimedia asset (e.g. color descriptors).*

Definition 3.9 (definition scheme). *Set of rules, relations and semantic relations between its parts either D or DS (e.g. spatio-temporal structure schemes).*

Definition 3.10 (description definition language). *Extended XML schema-based metadata definition language describing rules of how to create, extend and modify D and DS.*

Specific extensions of XML schema are necessary to cope with matrix and array data types, as well as built-in data types (e.g. `basicDuration` as derived data type from `string`). In principle, the DDL is a metadata definition language, describing how instantiated MPEG-7 documents are structured. DS and D are part of the MPEG-7 metadata definition language and describe dependencies within MPEG-7 documents (e.g. MPEG-7 root node) and concrete features of multimedia assets. Descriptors represent low-level (e.g. color) or high-level (e.g. semantic annotation of objects) features of multimedia assets [130].

Methods to obtain values for D or DS are non normative, and therefore are left open to the application developers. MPEG-7 metadata definitions are a normative part of the standard. Thus data representation is standardized and interoperable — the process of applying and obtaining it is not.

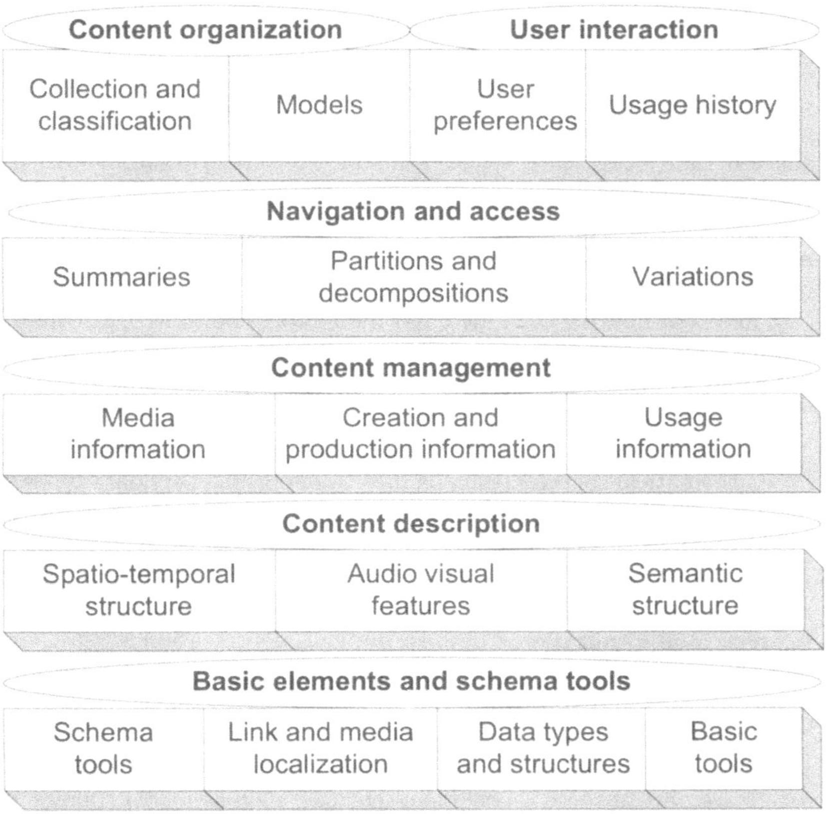

Fig. 3.6. An overview of MPEG-7 metadata definitions (based on [130])

MPEG-7 defines a set of tools categorizing D and DS according to their underlying functionality within the multimedia description schemes (MDS). MDS describe abstract structures, relations, components and sub components of MPEG-7 documents. Features of concrete multimedia assets are covered by visual and audio metadata definitions based on the MDS. Figure 3.6 gives an overview of the principal components of MPEG-7 metadata definitions.

3.4.3 Basic Elements and Schema Tools

Basic Elements define a set of general and valid metadata definitions as a basis for more abstract definitions of audio, video and other MDS descriptors. Several parts of MPEG-7 metadata definitions utilize basic elements as the origin of their newly defined descriptors. Basic tools consist of the following components [130, 117, 115].

- *Basic tools:* To build complex and more application specific metadata definitions, abstract definitions as a basic tool set have to be defined. These include textual annotations, relations between description schemes and graphs. Textual annotations are natural language constructs and less restricted than other data types (e.g. free text, description of people and places, keywords).
- *Schema tools:* Schema tools define the process for creating valid MPEG-7 metadata instances, structural elements and the management of metadata definitions. A base type hierarchy (e.g. MPEG-7 root node, types for D and DS) for encapsulating valid MPEG-7 metadata definitions is provided. Additional information, such as versioning, creational information and rights assignment enriches the overall spectrum of metadata definitions.
- *Data types and structures.* Schema tools provide information about the general layout of MPEG-7 documents. Additional data types, especially for mathematical constructs are defined by supplementary data types. This includes constructs for vector and matrix representation, strings (e.g. localization information) and statistical information (e.g. probability vectors).
- *Linking and localization tools.* Linking and localization of and within multimedia assets requires unique — either implicit or explicit — identification, reference and localization methods. Implicit identification requires unique identifiers (e.g. content identifiers, `UniqueID` data type, ID attribute for metadata structures). Explicit identification for media is supported via media locators (e.g. URI). Linking descriptions with e.g. XPath [40] and temporal aspects for referring to temporal segments of multimedia content assets are part of these tools. Time schemes, such as media time or world time representations are also part of these tools.

The creation of valid MPEG-7 documents can be based on either complete or partial descriptions, where only subsets of MPEG-7's metadata definitions are utilized. Partial descriptions are relevant for event-based document updates or simplified MPEG-7 document structures.

3.4.4 Annotating Multimedia Assets

Content description tools are devoted to the structural decomposition of multimedia assets, audio-visual definitions and to providing conceptual semantic information.

Structural tools allow a spatio-temporal and hierarchical decomposition of multimedia assets. Audio-visual segments/regions, moving regions, audio-visual regions/segments, spatio-temporal segments and its temporal (e.g. time overlaps, duration) and spatial structural (e.g. left of, right of, south of) relations are some examples of decomposition definitions [130, 117, 115].

Semantic or narrative meaning is described by events, objects, relationships or concepts connected by a network of links. Events (e.g. a happening) occur to objects (e.g. people) and are triggered according to abstract conceptual models (e.g. state diagram). Semantic relations therefore provide a powerful toolset to describe and create narratives. They enable telling stories with a meaning, world, background and context.

Example 3.11 (spatio-temporal structure of video content). The example presented in Fig. 9.3 represents the spatio-temporal structure of video content. Selected video segments — or video shots — contain a reference key frame. Information about the key frame representation and location is given. Key frames contain hyperlinks, annotated by utilizing still region descriptors with associated actions. Still regions are simple 2x2 boxes describing position, width and height of rectangles.

3.4.5 Grouping Multimedia Assets: Content Organization

The MPEG-7 metadata tools for content organization group or cluster multimedia assets by certain criteria to collections. Grouping criteria range from simple logical packaging structures (e.g. folder containing holiday images with their description) to more complex clustering models (e.g. movie genre classification). For this purpose two toolsets are available: collections and models [130, 117, 115].

Collections describe multimedia asset groups or collections, their subgroups and relations between them. Similarity criteria define relationships in groups or between groups of multimedia assets.

Models are used to associate characteristics with instantiated groups of multimedia assets in the form of probability, analytic, cluster or classification models. Statistical or probabilistic models act as classification criteria, describing the relation between features of instantiated multimedia assets (e.g. hidden Markov models).

Example 3.12 (subjective image quality estimation). A very simple estimation of image quality can be determined by its histogram distribution. Low quality manifests itself by a less clustered histogram distribution caused by lossy compression. High quality images show equally distributed histogram values. Content organization tools help to group these into high quality and low quality image clusters by applying simple probabilistic models.

3.4.6 Managing Conventional Media Archive Information

Conventional media archive information is defined within content management tools and describes the whole life-cycle of multimedia assets, from producer to consumer. Parts of the definition are media information, creation information and content usage information [130, 117, 115].

- *Media information* relates to the independent identification of unique multimedia assets and includes different media profiles. Examples include coding format, video resolution, transcoding descriptions, adaptation of multimedia assets to different network and terminal capabilities, etc.;
- *Creation and production information* holds data about the overall process of creating multimedia assets (e.g. actors, places), classification schemes for searching and filtering (e.g. genre, localization, language) and related materials for advanced services (e.g. additional resources, information sources, parental rating, reviews);
- *Content usage information* describes rights of usage (e.g. user rights and authorizations) and multimedia asset history (e.g. asset changes, availability, previous audience, financial information).

Example 3.13 (media information). In Fig. 9.3 description metadata includes conventional media archive information for the MPEG-7 document itself. Multimedia asset specific information, in our case for an overall video, is represented via media information, creation information and usage information within the scope of video descriptors. A key frame is separately annotated by similar information.

3.4.7 Easy Navigation and Access

Efficient modes for accessing and browsing through multimedia databases are addressed by navigation and access tools. They enable features such as summaries, partitions and decompositions, and variations follow a consumer-oriented way in facilitating large multimedia archives [130, 117, 115].

Summaries address the hierarchic decomposition of multimedia content assets for preview as well as simple text-based search facilities. Partitions and decompositions relate to spatial-temporal inner media asset navigation to gain access to specific features. Different views and modalities assist the user in adapting existing media material to local facilities, as well as user browsing habits (e.g. representation of an image in text or low resolution).

3.4.8 Personalization, User Interaction and Consumer Profiles

Large amounts of multimedia content require intelligent ways of consumer profiling and personalization. MPEG-7 supports user preferences to filter multimedia content and usage history. The latter enables future prediction of media usage, user categorization and profiling of consumers [130, 117, 115].

3.4.9 Audio Descriptors

A tool for audio descriptions is provided within MPEG-7 for any type of audio content features. This includes feature for music, speech, instruments and spoken content. Similarity measures are based on features such as timbre, melody, harmony and waveform [114, 130].

Table 3.5. MPEG-7 audio descriptors [114]

MPEG-7 Descriptor	Features
low-level descriptors	temporal and spectral descriptors grouped into basic, basic spectral, signal parameters, temporal timbral, spectral timbral and spectral basic descriptors
high-level descriptors	descriptors for sound recognition and indexing, spoken content descriptors, melody descriptors and musical instrument timbre descriptions
spoken content descriptors	speaker information, spoken content header, word and phone lexicon, spoken content lattice
sound classification and similarity	spectral basis functions, sound classification schemes, sound probability models, sound model histogram, sound model state path

A more detailed description of metadata definitions for audio content can be found in [114].

3.4.10 Visual Descriptors

MPEG-7 provides a visual descriptor tool for describing any type of visual content feature. This includes features of videos, images, 3D models, 2D models, etc. Descriptions can be used for measuring similarity and they act therefore as a basis for searching and filtering visual content. Basic MPEG-7 tools support localization and spatio-temporal descriptions [118, 130].

Basic, more abstract features (e.g. color) are defined by low-level descriptors. High-level and more complex descriptor schemes are present for specific application scenarios (e.g. person descriptors). Table 3.6 gives an overview of the complete visual descriptor set. A more complete account of specific visual metadata descriptions can be found in [118].

3.4.11 MPEG-7 Systems

The MPEG-7 system architecture specification defines an efficient way for streaming metadata in parallel to audio/video content to consumer terminals. It is the foundation for the realization of a metadata protocol stack model, including an optimized metadata representation [130, 116]. Figure 3.7 shows

Table 3.6. MPEG-7 visual descriptors [118]

MPEG-7 Descriptor	Features
color descriptors	color spaces (e.g. HSV), dominant color, scalable color, color structure, color layout, group-of-frames color, group-of-pictures color
texture descriptors	homogeneous texture, texture browsing, edge histogram
shape descriptors	region-based descriptors, contour-based descriptors, 3D shape
motion descriptors	motion trajectory, motion characterization, motion activity, camera motion, parametric motion

a schematic overview based on the ISO/OSI 7 layer reference model. Key features include:

- *abstract MPEG-7 system architecture model* including binary MPEG-7 format definitions;
- *optimized binary format for metadata (MPEG-7 BiM)* for partial and full metadata updates, low bandwidth consumption, small parsing times at consumer terminals, compression of MPEG-7 schema and document definitions and encryption;
- *support for various transmission modes* ranging from push and pull delivery models, packet-based metadata delivery, multi-protocol encapsulation, synchronization, to streaming facilities including binary storage;
- *bi-directional mapping* of binary and textual metadata definitions including syntax checks for document validity and well-formedness documents.

Binary representation enables synchronization of metadata trees between different devices by partial or complete tree updates. Metadata definition languages and metadata instances are binary encoded and decoded at end-devices. Optional encryption guarantees secure data delivery e.g. for personal or conditional access data. Validation and check for well-formedness maintain document integrity. MPEG-7 systems can be utilized to transmit several other XML-based metadata definitions.

The underlying principle is based on mapping complete or partial metadata trees into so-called *access units (AU)*. An AU consists of a header, containing an optional digital signature and a payload description. Each AU carries one or more *fragment update units (FUU)* in its payload. The header of an FUU holds data about navigation path, insertion modus and how the update has to be performed. Binary descriptions themselves are part of the payload of an FUU. See [130, 116] for further details.

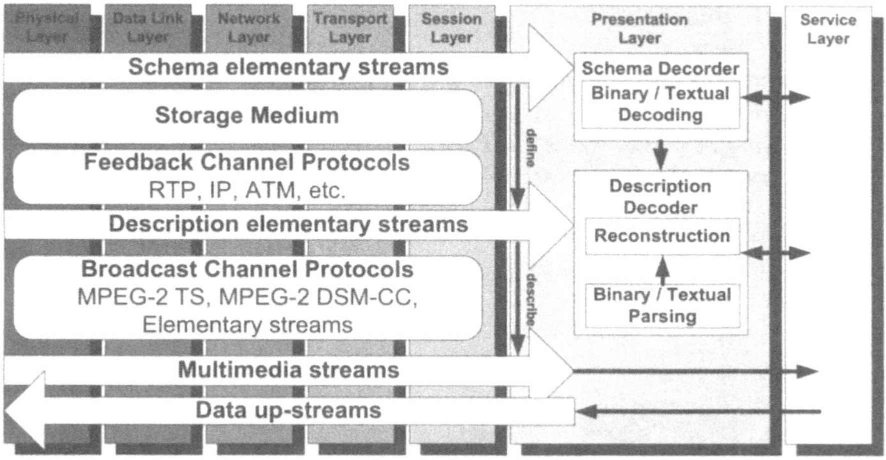

Fig. 3.7. MPEG-7 reference system architecture in the ISO/OSI reference model [17]

3.5 MPEG-21 Packages Multimedia Assets

The newest multimedia standard published by MPEG is MPEG-21, whose definitions differ from those of previous MPEG standards. The purpose is the transparent, protected and interoperable use of digital multimedia assets throughout their value-chain. The unit of interchange is a structured digital object (digital item) described by metadata and referencing to related multimedia content assets.

MPEG-21 standardization is still in process and requires additional efforts for the finalization of the overall standard family consisting of seven parts.

- Part 1 (*Vision, Technologies and Strategy*) [103] describes the MPEG-21 framework on a general level, introduces use-scenarios, sample applications and general conformance guidelines.
- The unit of exchange in a distributed MPEG-21 compliant environment is a *Digital Item (DI)*, which consists of a description part and references to multimedia content assets. DIs are defined in part 2 (*Digital Item Declaration*) [107].
- Part 3 (*Digital Item Identification and Description* [108] is dedicated to uniquely identifying and describing DIs. Each element of a DI has to be unique throughout the value-chain.
- To protect DIs from unauthorized use (e.g. access of music files by customers that copy them illegally), a generic IPMP framework is introduced in part 4 (*Intellectual Property Management and Protection (IPMP)*) [109] of the standard framework.
- Still, the introduction of DIs requires the definition of who is allowed to do what with them. Part 5 (*Rights Expression Language*) [110] and part

6 (*Rights Data Dictionary and MPEG-21*) [111] cover the rights management of users.

- The next specified and standardized part of MPEG-21 is part 7 (*Digital Item Adaptation*) [112].
- *Processing of digital items (DIP)* will be defined in [104];
- Other standardization efforts of MPEG currently in progress are the evaluation of persistent association tools in [105] and a resource delivery test bed in [106].

An example of a compiled DI is shown in Fig. 3.8.

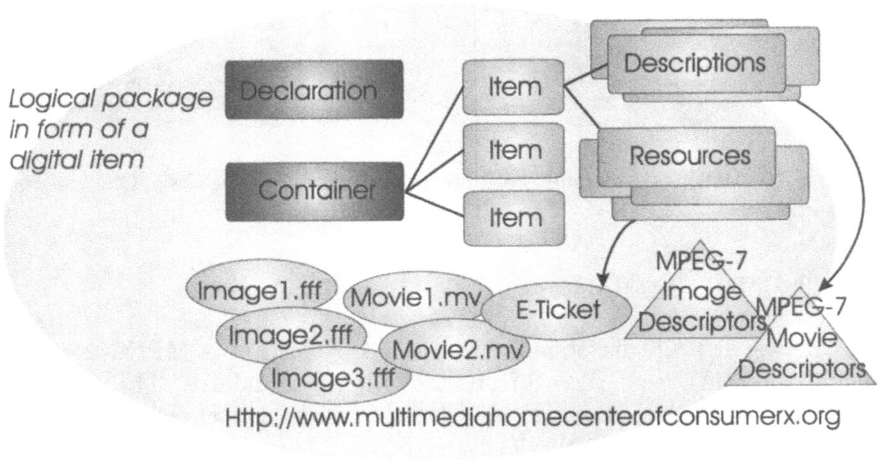

Fig. 3.8. MPEG-21 digital item use-scenario for compiling a digital holiday collection

In order to understand MPEG-21 it is important to consider what a DI is and on which principles it is based. In the digital world, DIs are the digital counterparts of physical "real" objects such as books. Resources associated with descriptors are the basic elements of a DI. A resource can be of any multimedia asset type (e.g. video, audio, metadata), where descriptors come mostly from other metadata standards and describe the resource (e.g. variations of resource, content, bandwidth requirements needed to consume the resource).

In addition to the representative elements, presentation and an underlying model of the DI are also of special interest. Each DI is based on a content description model (e.g. encoding technique, content characteristics) that can be presented in different ways (e.g. visualization of the logical structure of DIs with 3D graphics or displaying 3D graphics as content of DIs). Putting all characteristics and new potentials of MPEG-21 together, MPEG-21 is more than just a simple metadata standard. MPEG-21 is the "21st century multimedia framework" [26]. The following features are enabled by MPEG-21:

- *packaging and catalysis of existing multimedia assets* (content, resources and metadata) and their associated descriptions independently of their locations to DIs as an atomic unit for processing;
- *creation of more awareness of digital content* by introducing a perceptive object for the consumer, which can be modified, exchanged, altered and stored;
- *access and consumption of content* anytime, anywhere, anyhow and by any means by adapting, transcoding and transforming digital items according to network and terminal requirements;
- *interoperable digital rights management* as a factor for successful business models for distribution of any type of digital content (e.g. music files, video files);
- *unified content management* based on an abstract data structure for storage, processing and adaptation of multimedia assets;
- *metadata definitions* for logical digital packages, rights expression, digital rights management, file format, etc. allow heterogeneous and interoperable solutions for metadata-driven digital item deployment.

3.5.1 Perception of Multimedia Assets through DIs

Consumers are confronted with myriad megabytes of multimedia assets in arbitrary types of content: music files, digital camera images, videos, documents, etc. The consumer is confronted with multimedia asset management rather than enjoying and creating digital content. Physical "real" objects are more than just simple file names or bits and bytes: they can be perceived, touched and organized into shelves to create personal relationships.

Creation of personal binding to digital content is a rather hard task, if digital objects are perceived as single unorganized separate entities. Digitalization means easier creation of digital content represented in different types and distributed over various storage media (e.g. digital camera, PCs, PDAs). Au-grand creation of digital objects requires easy sorting out and organizational models for them. From the consumer perspective it is hard to create a "feeling" of content, when it is located on a memory stick or similar device. DIs assist in the creation of perception for digital objects on a higher level. Awareness, creation of a personal relation to digital content, feeling digital objects and placing emphasis on the creation of awareness for digital content are major concerns [128].

MPEG-21 and its concept of DI mean a possibility for creating a representation and presentation of digital content. A DI is perceived as a digital object entity — just like a "real object" (e.g. books). DIs can be put into a virtual shelf, one can arrange them nicely, annotate them with some tags describing their content or loan them to friends.

Example 3.14 (digital holiday memories). After two weeks of holidays on a Greek island multiple leisure videos, photographs and audio recordings are brought home in "digital luggage". Different storage formats, multiplication of digital materials and different annotation methods require consumer-friendly mechanisms to sort out, present content to family and friends and to archive materials semi-automatically. MPEG-21 supports digital archiving and integration of multiple metadata formats to one digital item. The consumer is able to compile a multimedia show of his holiday experiences and store it on the multimedia home center (see Fig. 3.8).

It is a fact that multi channel applications (editing once — publishing to various end-consumer devices) are required for economical and rapid distribution of multimedia assets. Once an application is created it can be transparently distributed over heterogeneous networks to ubiquitous consumer device technologies. Digital content must be accessible to the consumer anywhere, anyhow and anytime. MPEG-21 offers adaptation possibilities, with the help of which content can be created and later adapted to different network bandwidths, consumer devices, software architectures, etc.

A BSP could provide different variations of a TV program in parallel: low-bit-rate versions for broadcasts to mobile devices, high-bit-rate versions for transmitting content over a broadcasting infrastructure. First, a DI announces the provided variations of available TV programs. The consumer or consumer devices decide which one is the most adequate based on the available resources. Invisible to the consumer are transmission protocols, resource requirements, adaptation mechanisms for enjoying content on multiple end-devices, service contexts and algorithms for processing DIs.

3.5.2 Digital Item Declaration (DIDL)

To realize digital item-based services an abstract, coherent and self-containing data structure has to be defined. The DI represents a logical digital package assigned with a unique identifier and encapsulates multimedia content assets and their metadata definitions to a new logical atomic unit of exchange. MPEG-21 standard part 2 [107] handles this aspect. Table 3.7 shows metadata definitions related to the DI.

A digital item is self-containing and holds information about its content and references or encapsulates multimedia content assets in a new virtual data package. A *declaration part* defines descriptors and a metadata definition valid throughout the digital item for its description, whereas an *item* or *container* part groups several concrete instantiated resources, configurations and descriptors together.

Three different types are distinguished: *typed items* are digital items put in a specific perspective or applied within a certain domain; *contextual items* are purely holding descriptions and metadata definitions; *content items* are items that hold resources only [26].

Table 3.7. MPEG-21 digital item declaration (DIDL) metadata definitions [107]

Structures for Building DIs	
root node	top-level node for building digital items and parsing entry point
container	groups items
item	groups sub items and container to items
component	encapsulates a resource and assigns it with descriptors and conditions
resource	reference to a multimedia content asset
fragment	reference within a multimedia content asset
descriptor	contains descriptors in arbitrary metadata definition languages
anchor	encapsulates a fragment in a multimedia content asset and assigns it with descriptors and conditions
Configuration and Variation Parts	
choice	a choice descriptor groups selections and conditions and associates them with descriptors or reference descriptors
selection	selections manifest upon external inputs (e.g. user inputs) and associate actions and descriptors coming into force according to them
condition	creation of optional and required selections and their relationships within a selection or choice
assertion	predefinition of predicates and their values within a choice element
predicate	true, or false or undecided atomic logical statement
Auxiliary Definitions	
annotation	description and commenting purposes
statement	textual entries in a digital item; mostly used for inline raw XML code

Example 3.15 (digital item in broadcasting). An item for specific use in broadcasting assigned with a predefined set of descriptors is a typed item. To deliver pure resources, content items rely on this purpose only. An example is a video stream broadcast. Variation description (e.g. bandwidth requirements for multimedia services) encapsulated in a digital item relies on pure metadata definitions. Therefore they belong to contextual items.

When viewing the digital item declaration from the perspective of a metadata definition, three essential parts in building digital items can be distinguished: *structures for building DIs*, *configuration and variation parts* and *auxiliary definitions*. Structures for building DIs involve several components for building digital items and creating their shelf-like structure. Each resource is embedded into the DI via containers, relating multimedia content assets to its descriptors expressed under certain circumstances by configuration and variation parts of the digital item declaration.

Configuration and variation parts relate to compiling an item, its sub-items and containers according to certain conditional choices. This relates either to optional metadata descriptions, multimedia assets or to manifestation of

Table 3.8. Schematic file of an MPEG-21 digital item containing high-bit-rate and low-bit-rate variations of the same movie

```
<?xml version="1.0" encoding="UTF-8"?>
<DIDL xmlns:xsi=http://www.w3.org/2001/XMLSchema-instance
   xsi:noNamespaceSchemaLocation="....\MPEG-21-DI\DIDL.xsd">
   <DECLARATIONS>
      <DESCRIPTOR>
         ....
      </DESCRIPTOR>
   </DECLARATIONS>
   <ITEM ID="FITV_MOVIE">
      <CHOICE>
         <SELECTION SELECT_ID="LOW_BITRATE">
            <DESCRIPTOR>
               <STATEMENT>channels up to 500kbps</STATEMENT>
            </DESCRIPTOR>
         </SELECTION>
         <SELECTION SELECT_ID="HIGH_BITRATE">
            <DESCRIPTOR>
               <STATEMENT>channels over 500kbps</STATEMENT>
            </DESCRIPTOR>
         </SELECTION>
      </CHOICE>
      <COMPONENT>
         <CONDITION REQUIRE="HIGH_BITRATE"/>
         <REFERENCE URI="http://www.futureinteraction.tv/
           low_bit_rate_fitv_movie.avi"/>
      </COMPONENT>
      <COMPONENT>
         <CONDITION REQUIRE="LOW_BITRATE"/>
         <REFERENCE URI="http://www.futureinteraction.tv/
           high_bit_rate_fitv_movie.avi"/>
      </COMPONENT>
   </ITEM>
</DIDL>
```

required or optional parts of an instantiated DI. A digital item can contain descriptors for video streams encoded with different bit-rates meeting the capabilities of the available consumer resources (e.g. dependent on the network bandwidth available to the consumer, the consumer can download a low-bit-rate video from a specific location and optionally a high-bit-rate version from another location). Table 3.8 shows a schematic example of an instantiated digital item.

3.5.3 Digital Item Adaptation

Adaptation is a systematic approach of MPEG-21 through *digital item adaptation (DIA)* for matching available resources required for the distribution of multimedia assets.

Different application levels can be distinguished: on the *infrastructural level* adaptation is performed on core network equipment or the low-level hardware side (e.g. bandwidth-optimized broadcasting of video); on the *content level* adaptation mainly focuses on the transcoding of multimedia assets (e.g. transcoding of high-bit-rate video to low-bit-rate streams); and on the *service level* where whole services are created to be ready for multi channel distribution (e.g. automated customization of Web pages for digital TV and PC by adjusting font sizes and page layout).

Definition 3.16 (transcoding). *Alternation of coding and encoding formats for multimedia assets to other arbitrary types by an algorithm*

Definition 3.17 (transforming). *Alternation of metadata formats either of instantiated metadata, metadata languages or metadata definitions to other appropriate data structures.*

Definition 3.18 (adaptation). *Distribution of multimedia assets to various different types of consumer devices over arbitrary transmission media by customizing multimedia asset types to available resources via transcoding or transforming.*

Currently defined descriptors come from MPEG-7 standards, defining variations and the principle of content adaptation. UMA descriptors defined by MPEG-7 are categorized according to user characteristics, terminal capabilities, natural environment and network characteristics. UMA defines purely descriptors, whereas MPEG-21 extends the idea to the definition of a general adaptation engine [163].

The consumer can experience any type of content through any channel. An example would be broadcast shows sent in parallel over low-bit-rate channels to mobile phones and through high-bit-rate channels to commonly used digital TV equipment. Also overall top-down solutions e.g. for adapting network equipment including video streams to local facilities are potential examples for DIA [27].

The potential of adaptation also refers to making services accessible for people with special needs. For example, Jael Song has introduced color vision deficiency descriptors to make content available for color vision deficient people [161].

3.5.4 Road Ahead for MPEG-21

There is still more to be defined within the MPEG-21 standards. So far digital item fundamentals, adaptation and metadata definitions have been discussed.

Originally MPEG-21 was designed as a metadata definition for digital rights management. It is currently developing far beyond its original purpose and aims towards many more application areas than originally. MPEG-21 has more to offer in standards yet to be released:

- *intellectual property management and protection (IPMP)* will be defined in part 4 of MPEG-21 standards [109]. The central component is IPMP tools assigned with IPMP functionalities as e.g. decryption algorithms or smart card solutions. Tools can communicate with each other to exchange messages. Together with the rights expression language and the rights data dictionary, MPEG-21 provides a heterogeneous solution for dealing with IPMP issues;
- *rights expression language (REL)* will be defined in part 5 of MPEG-21 standards [110]. It assigns specific rights and permissions to digitally distributed multimedia assets. REL defines permissions and rights (e.g. printing, executing) on a very abstract level for machine readability, interoperability and augmented usage of multimedia assets through consumers. Examples for REL descriptions are usage permissions, usage conditions and authentication;
- *rights data dictionary (RDD)* will be defined in part 6 of MPEG-21 standards [111]. A dictionary containing entries for authorities, namespaces, expressions for REL, etc. is defined. It harmonizes and standardizes the structure for describing usage rights on a higher level;
- *unique digital item identification (DII)* described in part 3 of MPEG-21 standards [108] assigns unique content identifiers to items and their subcomponents.

Due to the fact that MPEG-21 is currently still evolving, future developments will focus on system architectural aspects with reference implementations. Another open point in MPEG-21 is the definition of a file format. This will also be covered in future standardization work.

3.6 MHP and Metadata

Metadata in MHP evolves around two features: definition of rigid metadata and provision of a whole metadata application environment. The basic rigid metadata is defined by standard broadcasting metadata, DVB-SI. The latter is XML-based metadata (granular metadata) including the standardization of a whole metadata application environment. MHP includes limited support for granular metadata through DVB-HTML. Current versions of MHP standards focus rather poorly on new metadata definitions. However, adding emerging metadata definitions into metadata-based MHP applications allows a wide use of granular metadata in digital TV. This leaves two ways for adapting metadata definitions on MHP compliant platforms: first, usage of metadata definitions as described in MHP standards [68] is based on rigid metadata; second, adding emerging metadata into applications built on top of MHP.

In principle, sub-setting and using existing metadata definitions that have to be built on top of the MHP application interface are dominant. For further reading we would like to point to [140]. A simple classification categorizes the used metadata definitions in MHP [68] according to their operative purpose:

- *provision and support for metadata processing:* XML version 1.0 as defined by W3C is based on SGML and has been adopted by DVB as the principal metadata definition. Of special interest is the DTD due to its simpler structure implying less performance consumption and faster parsing. Another standardized metadata language is DVB-HTML, based on W3C's XHTML, and especially designed for representing HTML content on digital TV devices (e.g. bigger font sizes, digital TV compatible fonts);
- *configuration and parameterization:* XML is especially interesting for configuration and parameterization management. MHP therefore uses instantiated DTD files for setting permissions (e.g. file permissions) and application security. This protects the consumer from unauthorized access and use of her data;
- *metadata application environment:* DVB-HTML is a bit more in DVB than just simple HTML, it is a whole XML application environment for metadata processing. This covers several components such as application life-cycle, permission settings for applications, execution rights, encapsulation of applications and their content in feedback and broadcast channel, etc.;
- *base platform for metadata parsing, representation and presentation:* DVB with its Java platform supports metadata parser development, but is limited towards predefined modules for doing this task. Exceptions are DOM definitions for DVB-HTML which provide dynamic metadata processing. Applications require parsers e.g. for rendering SVG-based graphical content on-screen. Built-in parsers would ease metadata-oriented programming. Future versions of MHP standards will address this issue.

MHP focuses on the standardization of application interfaces and metadata definitions for compliant consumer devices. Several definitions and use-scenarios of metadata belong to the consumption and interaction phase in the metadata life-cycle model. This implies including metadata processing facilities in MHP-compliant consumer devices.

3.6.1 "Metadata Way" of MHP

An overview of MHP related metadata issues is given in the following:

- *metadata definitions* relate mostly to adapted definitions from the W3C for broadcast usage. Also unique identifiers and locators for service retrieval are defined. Examples are DOM, XML, ECMA scripting, CSS and

XHTML. A basic structure for rigid metadata is inherited from basic DVB broadcasting standards and also belongs to MHP metadata definitions;
- *system aspects* relate to lower levels of transmitting and decoding of metadata. Especially rigidly annotated metadata is processed at this level. A protocol for conveying granular metadata (transmission, encapsulation, encoding and decoding) has to be defined on the system level. This includes metadata application encapsulation in either a broadcast or feedback channel network and also includes service accompanying metadata definitions. The delivery and processing of DVB-HTML applications are similar to those of an DVB-J Xlet (e.g. application control via flags such as autostart);
- *supported multimedia asset types* range from types known in HTML to be digital TV specific content types. They also include MPEG-2 A/V, MIME types (e.g. images, video), simple graphics, textual representations, Xlets, XML, CSS, applets, scripts, etc.;
- *application environment support* for DVB-HTML applications throughout the life-cycle phases (encapsulation — delivery — decoding — performing — stopping). MHP defines and standardizes the way these are performed in addition to other typical application environmental issues (e.g. application resource management, memory management, ...);
- *application programming interface* support for rigid metadata types. The DVB-J platform includes basic components for parsing service information. Processing granular metadata requires the implementation of additional modules on top of the existing DVB-J platform.

The ongoing standardization efforts of DVB will provide more metadata involvement and other extensions.

3.6.2 DVB-HTML

DVB-HTML is in principle HTML with digital TV related extensions and restrictions, restricted in the sense that digital TV equipment does not provide the whole range of capabilities that PC-based Web browsers do (e.g. screen size, font size, colors, font sets).

But DVB-HTML is also a bit more than a simple HTML-based metadata language. It is a complete application environment, with its own life-cycle and application encapsulation methods either for feedback or for broadcast channel networks. It can be seen as a new type of XML application with textual representation of metadata definitions.

Dynamic aspects and control mechanisms can be categorized as follows:

- *signaling and DVB-HTML application delivery:* encapsulation and updating of delivery descriptors within a broadcast data stream. Signaling means to add descriptors of applications into the broadcast stream. Other examples are application parameters, application entry points (e.g.

./apps/index.html), etc. Dynamic updates are further specified within the scope of this section;

- *application life-cycle* is controlled via an application environment and controlled via agents. It is similar to the one of Xlets. The dynamic behavior of applications can either be altered through mechanisms on the box or through external resources as explained in the following sections;
- *dynamic control of application behavior* through external signaling, consumer interactions or as defined by the application dynamic itself. Signaling allows altering the application life-cycle or event-based dynamic updates over the broadcast channel. Other examples are DOM events, synchronization of trigger mechanisms, etc. User interactions relate to browsing and interacting with the presented content and performing actions upon them. The structure of documents includes dynamic elements in the form of scripts, cookies, Xlets, applets, etc.

Applications are sent and encapsulated in broadcast and feedback network protocol suites. They are decoded at the consumer device. DVB-HTML applications are embedded within a user agent performing the life-cycle steps. The life-cycle of each application is similar to the one of an Xlet. Each page is rendered on-screen including its assigned multimedia assets. The consumer can navigate through the pages. External events either in the form of DOM events or application life-cycle modifications generated by outside resources are taken into consideration. They enable dynamic updating or influencing the transitions between the states of the application.

3.7 TV-Anytime

Newport Beach, California, USA was the place where the TV-Anytime forum was founded in 1999. The creation of an open public standard applicable in postproduction, distribution, consumption and interaction of digital TV content was the entire goal. "Application on digital TV equipment shall enable applications to exploit local persistent storage, involve network independent delivery, interoperable and integrate in existing systems and specify necessary security structures" [81].

Four working groups contribute to TV-Anytime: each of them is focused on different aspects of standardization processes: business models; system, transport interfaces and content referencing; metadata; and rights management and protection. MPEG-7 based on XML has been selected as the source for metadata definitions. Therefore TV-Anytime integrates well with MPEG-7 standardization efforts.

TV-Anytime is based on the following features and characteristics:

- consumer device-oriented metadata definition for metadata exchange among SP, BSP, CMHN with or without feedback channel;

- wide acceptance in common broadcasting related standardization processes (e.g. DVB, ETSI, EBU) including reference implementations and core experiments;
- TV-Anytime phases (publish, search, select, locate, acquire, view and finish) enable easy integration and embedding in existing digital broadcast system work-flow models;
- applicable either as a digital TV-focused standard or as a basic toolset for more advanced metadata schemes;
- standards cover broadcast channel and feedback channel network, secure transmission, binary metadata delivery, client system architecture, rights management, etc.

The TV-Anytime Forum has produced a set of specifications, called the specification series:

- S-1, "Benchmarking Features" [86] handles business model related aspects;
- S-2, "System Description" [87], contains the TV-Anytime principles and system descriptions along with several use-scenarios and examples. Both, S-1 and S-2, are informative;
- S-3 is normative, "Metadata" [88], defining the metadata descriptions of compliant systems;
- S-4, "Content Referencing" [89] is also normative and contains the specification of local resolution, localization and content acquisition;
- S-5 is currently not published, but covers rights management and protection;
- S-6, "Metadata Services over a Bi-directional Network" [84], standardizes binary transmission and service transport and service descriptors;
- Metadata protection mechanisms in bi-directional networks are normative and standardized in S-7, "Bi-directional Metadata Delivery Protection" [85].

3.7.1 Personal Data Recorder

From the functionality point of view, TV-Anytime is a consumer device-oriented metadata standard. Its simple goal is adding personal data recorder functionality to consumer digital TV equipment.

Figure 3.9. shows the dynamic behavior of TV-Anytime. Metadata definitions for automating and personalizing recording processes are defined. This includes content descriptions (e.g. program reviews), consumer related data (e.g. preferences, usage history), instance description (e.g. program location), etc. More generally, TV-Anytime and its metadata definitions allow "the consumer to find, navigate and manage content from a variety of internal and external sources including, for example, enhanced broadcast, interactive TV, Internet and local storage" [88].

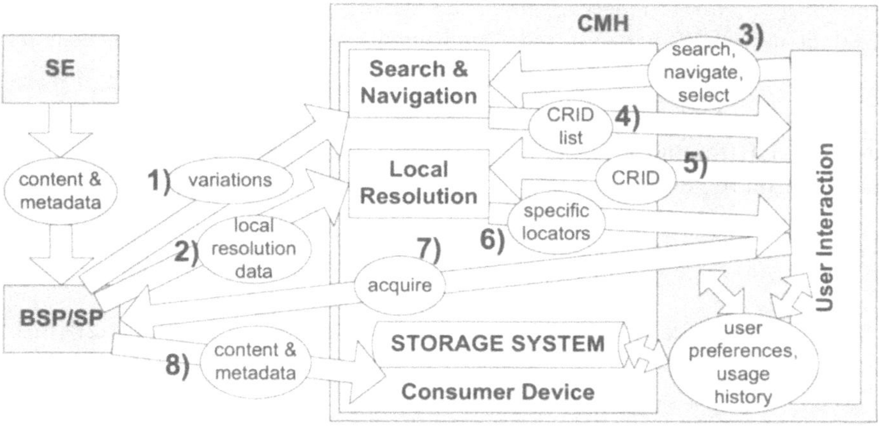

Fig. 3.9. Dynamic behavior of a TV-Anytime system as described in TV-Anytime standards for acquiring content as shown in [88]. Steps 1–2 belong to publishing, steps 3–4 to search and select, steps 5–6 to content location and steps 7–8 to acquire, view and finish

3.7.2 Content Reference Identifier (CRID)

The purpose of content referencing is to provide the consumer with a mechanism to refer to multimedia assets or multimedia asset subcomponents independently of time and location [89]. The consumer is capable of searching, navigating and selecting services with specific characteristics (e.g. title, language) without specifying location or when the service is available. Actions associated with the selection are triggered (e.g. recording of movie) when the multimedia asset is available on-line.

To understand the process of content referencing three different identifier types are defined [89]:

- *CRID* explicitly references multimedia assets or multimedia asset subcomponents. It is time and location independent and supports the selection process. CRIDs can be resolved to other CRIDs, specific locators or identifiers. It consists of two parts: an authority following the syntax defined in [89] and a data part. The data part is compliant with URIs [23] and can be any freely defined string;
- *locators* are location and time dependent for multimedia asset instances. After the selection process CRIDs are resolved to locators, which enable the consumer to acquire services. A locator also consists of two parts. Its delivery protocol dependence is manifested in the first part describing the transport protocol. The second part is URI-compliant [23]. Examples include DVB locators, holding digital TV program locations and sending time;

- *identifiers* are uniquely assigned to CRIDs and represent a location dependent identifier for instantiated multimedia assets. The first identifier part is a valid Internet domain name [149] followed by a URI-compliant string.

Steps 2 to 7 in Fig. 3.9. show the process from search to content acquisition and how CRIDs and locators are utilized. Local resolution either performed locally or on remote servers is the process of interpreting CRIDs or identifiers. Rights management or access rights mechanisms are associated with content resolution and acquisition.

Table 3.9. Syntax of CRIDs, locators and identifier as defined by TV-Anytime

Syntax	
CRID	`CRID://<authority>/<data>`
locator	`<delivery protocol>:<delivery protocol attributes>`
identifier	`imi:<registered internet domain name>/<data>`
Examples	
CRID	`crid://futureinteractiontv.org/channel1`
locator examples	`dvb://123.123.123` or `ftp://ftp.futureinteractiontv.org/f/mov.mov`
identifier	`imi:def.com/1`

3.7.3 Metadata Process Model

Normative metadata definitions are standardized in [88]. The TV-Anytime data model separates content and metadata flows. It focuses on the horizontal and vertical integration of its definitions during postproduction, delivery and consumption:

- *content creation:* during content creation metadata from different sources is collected, aggregated and edited to obey a consumer-oriented form of metadata representation. Aggregation means to harmonize metadata from multiple service providers and to transform it to a suitable form;
- *delivery:* delivering content and metadata to consumers including the definition of a program schedule and related information is a part of this phase. Local resolution, involving search and navigation mechanisms and transmission of data is part of this phase;
- *consumption:* consumer-oriented functionalities, such as consumer preferences, acting as an interface between the consumer and technology and data storage convolve the functionality during this phase.

3.7.4 Metadata Definitions

Specification series S-2, System Description [87] shows examples and use-cases and provides a top-down view of TV-Anytime metadata definitions.

Normative metadata definitions are specified in specification series S-3 [88] of TV-Anytime standards. The MPEG-7 DDL based on XML schema has been adopted as the principal metadata definition language. A unique namespace distinguishes TV-Anytime metadata definitions from others: (`xmlns:tva="urn:tva:metadata:2002"`).

TV-Anytime metadata categories with their descriptor sets are defined within its standards. Four categories of metadata definitions are present for different purposes [87, 88] and are shown in Table 3.10.

Table 3.10. TV-Anytime metadata categories

Descriptor Types	Example
Content Description Metadata Category	
basic	CRIDType, TVAIDType
descriptive	title, synopsis, genre
audio/video information	file format, A/V attributes, bit-rate
program information	type of variation
group information	episodes of a soap opera
media review descriptors	rating, newspaper review of a movie
optional metadata definitions	mandatory and required metadata
Instance Description Metadata	
program location entities	program schedule (whole EPG)
program location	one EPG entry
service information	TV channel, Channel 1 — Science
Consumer Metadata	
usage history and user preferences	implicitly or explicitly collected data from or about the user (user watches news every day at 9:00)
Segmentation Metadata	
basic segment description	segment title or synopsis
segment information	temporal location of a segment
segment group information	compilation of segments
segment information table	table of all segments metadata descr.

Profiling of goal platforms to adapt the rather complex TV-Anytime descriptors to different local hardware facilities is guaranteed through mandatory and optional entries in final TV-Anytime files. This enables consumption of media content throughout a wide variety of end-user devices, from low-performance to high-end PC-powered systems.

Content Description Metadata

The purpose of this metadata category is the description of whole programs including variations and basic requirements defined in [88]: variations relate either to different program genres (e.g. adult/children movies), subdivisions for publication (e.g. five movie parts), aggregations of programs (e.g. form multiple creators), program series (e.g. episodes), collections of series (e.g. summer hits) and attributions (e.g. movie tribute).

Besides simply grouping program content, a wide variety of typical broadcast relevant descriptors are defined to obey a coherent solution for service to search, locate, select and personalize digital TV services. This includes content descriptors (e.g. genre, parental guidelines) that ought to be matched with user profiles or previous user watching habits.

Instance Description Metadata

Final play-out versions delivered to the consumer (instantiated media material) need to be annotated with localization mechanisms and mechanisms for announcing services (e.g. EPG). Metadata definitions have the purpose of locating and determining scheduled items and describing provided service types and services.

Consumer Metadata

Consumer metadata defines metadata structures for defining user identification, user group identification, user profiles and user history. The usage history is based on ISO/IEC 15938-5 [115] (MPEG-7 MDS). Consumer metadata enables a very wide range of novel application scenarios [88] such as content and usage tracking, personalization, exchanging viewing histories between business value partners and payment models.

TV-Anytime enables monitoring several actions undertaken by the consumer: search, filtering, selection, browsing and viewing. The granularity of the recorded actions depends on the specific application and content type. A huge set of descriptors enables BSPs or SPs to profile consumers and consumer groups e.g. for personalized advertisements. The consumer device can record automatically digital TV content-based on consumer profiles. Data mining and personalization algorithms are not part of TV-Anytime definitions, but will find their application for different scenarios in digital TV.

Segmentation Metadata

This type of metadata description associates metadata within A/V streams. Each complete A/V stream is divided into segments or groups of segments. Each segment or segment group can be tagged with specific metadata definitions. Collection and grouping of segments is not only related to in-stream segments. Grouping and building metadata descriptions of any type of segments is possible. Several segments are collected in a segment information table, holding all program and program overlapping segment metadata descriptions.

Example 3.19 (news broadcast). Each complete news broadcast is divided into a sequence of news contributions or sequences. Each contribution represents a news segment. TV-Anytime enables the collection of different sequences of news segments of a certain topic. A potential scenario would be to match news segments to usage history descriptors or user profiles for automatic recording. A set-top box could automatically collect news segments of different BSPs during the day and record them.

3.7.5 Broadcast Channel Aspects

A push environment requires different encoding and encapsulation mechanisms as defined in [88]:

- *metadata representation in delivery* enables partial or complete update of TV-Anytime metadata between devices encoded by MPEG-7 BiM [116];
- *metadata delivery* encapsulates and transmits metadata between devices and is highly dependent on the delivery protocol. TV-Anytime descriptors relate to the description of the delivery process;
- *indexing* refers to an intelligent database system to navigate through metadata pieces.

3.7.6 Feedback Channel Aspects

TV-Anytime also devotes its standards to systems equipped with feedback channel network capability [84, 85]. This is especially of interest for data and service exchange between parties in the value-chain, such as BSPs, SPs, consumers or within a CHMN. Following additional capabilities over an "all-IP" network' are enabled [84, 85]:

- *service discovery* enables yellow-page-like lookup of services based on commonly known Internet standards such as UDDI and WSDL [170] explained in [60];
- *service adaptation* harmonizes the variety of metadata exchanged between devices, due to the different client and server capabilities of providing or consuming metadata;
- *metadata exchange* point-to-point or in request-response between partners in the value-chain over an "all-IP" network (e.g. user profile exchange within CHN devices);
- *delivery protection* is devoted to secure exchange of metadata-based on the secure socket layer (SSL) or transport layer security (TLS) protocol [54].

3.8 SMPTE Metadata Definitions

The *Society of Motion Picture and Television Engineers (SMPTE)* defines a set of purely professional broadcast focused standards, engineering guidelines and recommended practices [74]. This chapter acts as a reference point and introduction towards further reading, as a complete description of the whole standard family would be beyond the scope of this book. We focus on elementary concepts, basic principles and describe concepts being part of the other metadata standards (e.g. MXF utilizes key-length-value coding). Many other standards such as AAF, GXF or MXF, explained later within this chapter, have been delivered to SMPTE for approval.

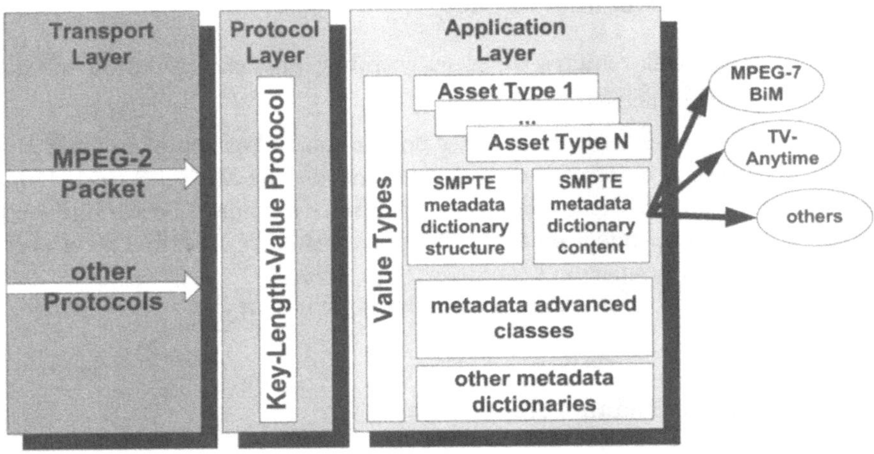

Fig. 3.10. SMPTE metadata related standards, guidelines and practices (adapted from [74])

3.8.1 SMPTE Metadata Dictionary (Content and Structure)

A metadata dictionary is a reference list for metadata definitions used in professional digital TV. SMPTE standardization efforts are dedicated to the standardization of the structure of the metadata dictionary itself within SMPTE 335M-2001 [76]. Besides a continuous registration of new metadata definitions utilized in digital TV, updating the dictionary entries is required. SMPTE RP210.4 [79] contains the overall content of registered metadata definitions.

Dictionary Version at Introduction	SMPTE Designator (first 8 octets	SMPTE Item Designator (last 8 octets following Metadata Dictionary Universal Label)	Data Element Name	Data Element Definition	Type	Value Length	Defining Document	...
5	060E2B3401010105	0301022001030000	XML document text	An XML document in MPEG-7 BiM form	Bytestream	Variable	ISO15938-1 section 7	...

Fig. 3.11. Metadata dictionary entry for MPEG-7 BiM as defined by SMPTE RP210.4 (from [79])

Initially 15 defined classes (of 127 possible ones) formed the basis of metadata classes described by the metadata dictionary structure [76]: identification and location, administration, interpretive, parametric, process, relational, spatio-temporal, organizationally registered for public use, organizationally registered as private and experimental. Besides metadata classes SMPTE 335M defines a dictionary element structure and format (e.g. value types, value length, value range) and dictionary maintenance issues (e.g. versioning, management, compatibility).

Example 3.20 (SMPTE metadata dictionary entry for MPEG-7). Figure 3.11 shows an example SMPTE metadata dictionary content entry in SMPTE RP210.4 for binary representation for metadata in MPEG-7.

3.8.2 Universal Material Identifier (UMID)

The universal material identifier (UMID) standard SMPTE 330M-2000 [75] globally assigns a unique identifier for multimedia assets of any type. A central registration office is responsible for keeping track of material numbers. Each UIMD describes multimedia assets either based on a content unit (e.g. one frame) or as clips (e.g. multiple frames) representing an integer number of a sequence of content units.

Figure 3.12 shows the structure for both types of defined UMID types: the basic UMID (32 byte length) simply identifies a clip without referencing its

3.8.3 Key-Length-Value (KLV)

SMPTE defines a data encoding protocol using a *key-length-value (KLV)* capable of containing data compilations or items in [77]. KLV is a very simple protocol based on a unique content key identifier, data lengths and the payload holding the data themselves as presented in Figure 3.13. The key identifier is based on a universal label (UL) specified in SMPTE 298M [77] providing information about the content type encapsulated in each KLV packet payload (e.g. metadata dictionary, version number). The length field is encoded as defined by ISO/IEC 8825-1 by the basic encoding rules (BER). The payload field contains the actual value of the data.

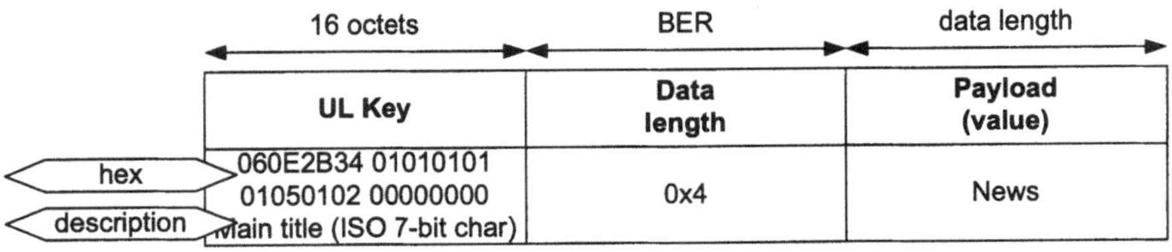

Fig. 3.13. KLV encoded packet with an example encoded value as defined in [77]

3.9 Advanced Authoring Format (AAF)

The Advanced Authoring Format Association (AAF) [14] is an independent non profit organization established in 2000. It was formed to enable the interchange of multimedia assets between content authors. Excellent descriptions can be found in [14, 15, 91, 10, 92], building the foundations for this chapter.

The goal of AAF is to obey a consistent workflow model to interchange, acquire, edit and preview digital content during postproduction phases. A standardized way for the interchange of multimedia assets during postproduction processes, such as editing, digitalization, online and offline rendering, content management, TV studios, multi track mixing, etc. is defined in AAF standards.

AAF is built on the following basic principles and functionalities:

- *data encapsulation:* wrapping of metadata, video, audio and data to content packages accessible throughout the overall life-cycle to obey a consistent data interchange;
- *support of broadcast workflow:* support of the broadcast workflow by tracking facilities, previewing of content and re-usage of compositions and subcomponents across platforms;
- *open standard:* openness towards typical broadcast editing and authoring tools, new content formats and data interchange protocols as well as towards standardization processes;

- *cross-platform:* standardization of a cross-platform software framework, object-oriented data representation, byte-stream structure and registration of new metadata formats;
- *editing:* editing and composing of simple content packages by involving plug-ins (e.g. effect filter), multi-track mixers, synchronization of subcomponents, timeline mapping, offline and 3D rendering, digitalization, etc.

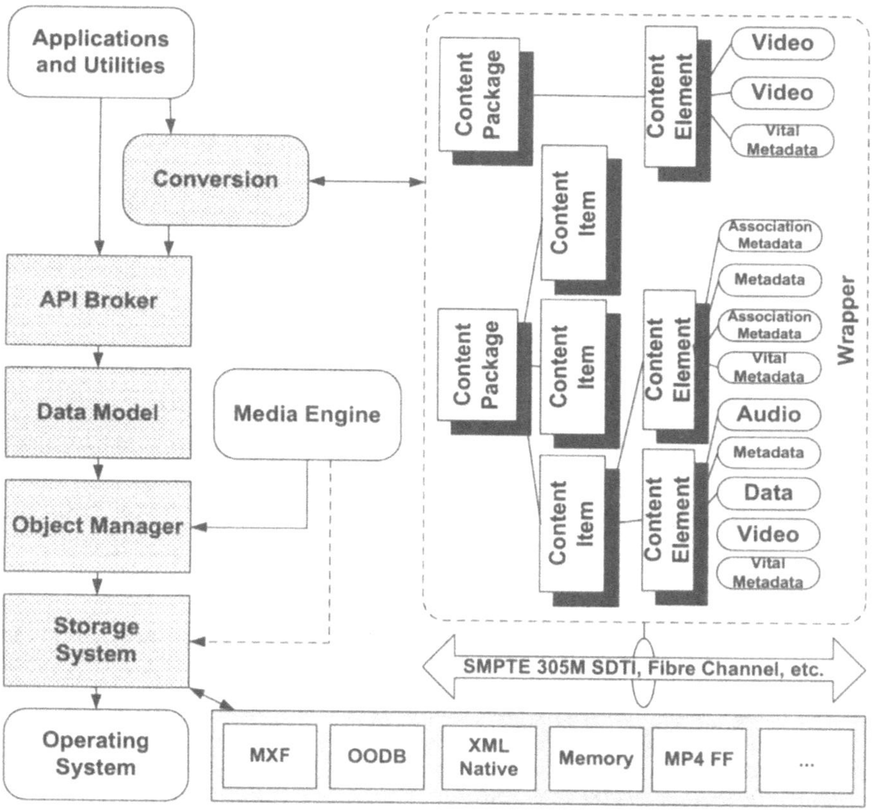

Fig. 3.14. The AAF architecture (from [167])

The high level of industrial support for AAF guarantees an open and consistent approach to standardizing broadcast environments focusing on APIs, utilities, data model, byte stream and registration. It defines neither streaming nor editing — it only focuses on data interchange between authoring tools. It mainly focuses on the data interchange and on the provision of a common work-flow model by integrating existing standards (e.g. MXF, SMPTE, XML).

Figure 3.14 shows the AAF architecture with its three metadata types: metadata structures for generic metadata, vital metadata for the description of multimedia asset types and association metadata to describe temporal compositions.

3.10 General Exchange Format (GXF)

GXF is standardized by SMPTE in SMPTE 360M [78]. It is excellently described in [59]. Both references form the basic source for this chapter. In principle, the *general exchange format (GXF)* belongs to the rigid annotated metadata definitions. Metadata definitions are restricted to describe the uncompressed or compressed content carried within a byte stream, and describing the overall byte stream multiplex structure. Due to completeness, its complementary character towards MXF and broad use in broadcasting systems, the standard is overviewed here.

GXF is an extended and more broadcast-focused version of the FTP protocol encapsulated in TCP and IP. It is designed for faster than real-time transfers in broadcast LANs and specifically designed for data exchange between archives, editing, source capture and play-out.

GXF is based on following principles and characteristics:

- *content representation:* GXF carries multiplexed and numbered tracks of video, audio and time codes in its payload. Each track consists of one or more media segments, i.e. video, audio or time code segments. Two types are defined: simple clips, where each track relates to a media segment from one media source and compound clips, where each track consists of one or more media segments;
- *packet-oriented transmission protocol:* as a packet-oriented protocol, GXF consists of a 16 byte header and variable payload. The header contains the packet type (e.g. end of stream) and the packet length. Packet types range from simple payload values descriptors (e.g. media types), track description tags, MPEG auxiliary information (e.g. bit-rate), to *unified material format (UMF)* description;
- *metadata definitions:* metadata definitions relate purely to the descriptions of stream contents and the transport multiplex. Descriptions of the stream are contained in the header of each GXF packet. UMF descriptions are especially relevant and focus on material description (e.g. material is open and shared), track description, user data and media description (e.g. chroma format).

GXF is especially designed for broadcast use and integrates successfully implemented concepts in broadcasting [59]. On-the-fly creation of GXF streams and encapsulation of streams on the source and destination server, based on open and standardized network protocols and encapsulation of broadcast typical compression formats (e.g. MPEG), integrates with other metadata families (e.g. MXF) and contains data to support archiving (e.g. lookup tables).

3.11 Material eXchange Format (MXF)

The *material eXchange format (MXF)* has been designed as a cooperative effort of the *Professional MPEG Forum (Pro-MPEG)* [150], EBU and the AAF. For further reading we suggest [21, 13, 31, 52] upon which this chapter is based.

MXF represents a stand-alone, self-contained file format from preproduction to postproduction and is substitutable by other production step-related formats (e.g. in postproduction by AAF). System integration for the purpose of interchanging digital TV content in an operational setting is guaranteed by embedding MXF into the overall broadcast workflow seamlessly to the user. Compression techniques, content asset conversion, storage access and transportation of different content asset types are invisible and work in the background.

The features, functionality and principles of MXF are:

- *system architecture:* MXF is built on top of standard IT equipment and existing broadcast equipment (e.g. streaming servers, content management systems and archives);
- *metadata definitions:* metadata defines multiple different features relevant in broadcasting, such as file structure, subtitles, teletext pages, creation location, time and date information etc.;
- *open and integrative standard:* MXF is an open standard and integrates existing broadcast related standards such as MPEG, SDI, SDTI, AAF, SMPTE 336M and DV. It is compression technique independent and applicable on top of existing broadcast equipment;
- *versatile file format [52]:* storage of compilations, streamable storage including view-while-transferring capability, wrapping of synchronization information and any multimedia content format in compressed and non-compressed form and embedding of higher layer information (e.g. for editing tools) are supported;
- *data encapsulation:* wrapping, storage and streaming are based on self-contained packages of complete multimedia assets without external references in the form of a versatile file format.

The very simple file structure of MXF consists of header, payload and footer information containing a multimedia content asset container (in MXF called an essence container) in its payload as shown in Fig. 3.15. Each multimedia content asset container consists of *material packages (MPs)* relevant for describing the output timeline, *file packages (FPs)* containing the description of multimedia content assets and *source packages (SPs)* containing edited derivatives of multimedia content assets as part of FPs. A number of tracks consisting of various segments contain whole, subcomponents or units of created multimedia assets. An overview is given in Fig. 3.15.

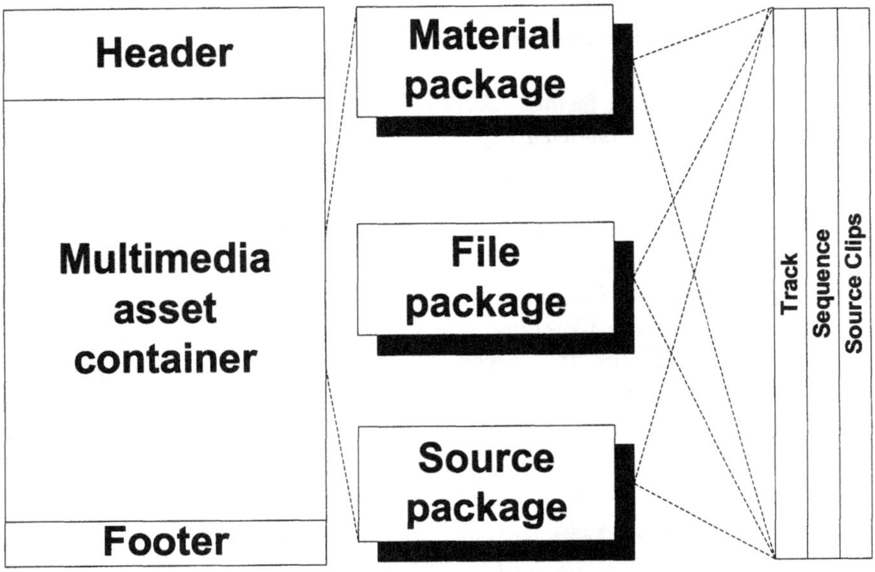

Fig. 3.15. Versatile file format in MXF consisting of different package types as presented in [52] and in MXF standards

3.12 EBU's P/META Metadata Exchange Scheme

In 1999 the *European Broadcasting Union (EBU)* created a project group called *P/Meta* to create integrated solutions for metadata-based broadcasting systems. SMPTE approaches the problem top-down by embedding newly emerging metadata definitions into its own standards. EBU [57] approaches the definition and use of metadata in a bottom-up fashion. It focuses its efforts on creating a new standard, the *P/Meta Schema* [100]. It strongly integrates and embeds existing solutions such as TV-Anytime, BBC's SMERF [20] data model for the exchange of metadata broadcasts between B2B business partners.

The outcome of P/META is a metadata definition, the EBU P/META Metadata Exchange Scheme [58]. It assigns descriptors especially designed for multimedia asset interchange among production, delivery and a multimedia content repository. Demonstrators show how multimedia assets can be interchanged among the content creator, the distributor, the archive manager and the consumer over B2B or B2C layers. Specific features of the EBU P/META Metadata Exchange Schema are [99, 100, 73]:

- *metadata definition:* syntax and semantic definition of a set of attributes, transaction sets, list of reference data (e.g. values), syntax and notation for construction [100, 99];
- *highly integrative:* high integration and facilitation of existing broadcasting-focused metadata definitions such as TV-Anytime, MXF, SMPTE Metadata Dictionary, EBU P/FRA (Future of Radio Archives) [63] and the Dublin Core initiative [47]. P/META resides on higher layers and does not define transmission, storage or exchange formats and purely focuses on the exchange of metadata;
- *consistent workflow model:* consistent workflow model for metadata exchange among third parties, content creators, archive manager, distributors, consumers and rights clearing authorities [99, 100, 73];
- *metadata language independent and system independent:* P/META is metadata language independent and applicable in typical broadcast systems;
- *open and non-profit oriented:* EBU defines open and non-profit oriented standards for the general usage.

The *EBU Metadata Exchange Scheme* [58] defines a common set of descriptors to exchange metadata definitions among professional broadcast value-chain partners. It is protocol independent on the transmission protocol layer, thus instantiated metadata definition schemes can be encapsulated in various broadcast specific protocols, such as KLV.

An atomic metadata unit is represented via attributes with an assigned name and identifier. Each attribute contains values of predefined types (e.g. Boolean, types of certain formats, types based on standardized value lists). Sets group and combine other sets or attributes to logical units of exchange (e.g. program descriptions). Name conventions ensure proper utilization of descriptors, as shown in the instantiated metadata exchange schema in Table 3.11. Each set starts with "s" and attributes begin with an "a".

3.13 Converging Broadcasting Metadata Standards

It is a rather difficult task to relate the overviewed metadata standards to each other due to their diversity and differences in application areas. An intelligent categorization divides them into broadcast specific metadata standards, multimedia metadata standards and generic metadata types. Figure 3.16 illustrates rigid metadata and granular metadata standards according to this classification.

Broadcast metadata standards have been developed purely for professional broadcast use. They are designed for reliability and for supporting the high demands of broadcasting. Most multimedia metadata standards have emerged from Internet-based service types. Both broadcast metadata and multimedia metadata are based on very generic metadata types mostly developed by the W3C (e.g. XML or HTML).

Table 3.11. Instantiated metadata definition schema as created on the basis of EBU Tech 3295

```
<?xml version="1.0" encoding="UTF-8"?>
<PMeta xmlns=http://www.ebu.ch/P_META
xmlns:erd=http://www.ebu.ch/P_META/ExternalReferenceData
xmlns:xsi=http://www.w3.org/2001/XMLSchema-instance
xsi:schemaLocation="http://www.ebu.ch/P_META.../P_META.xsd">
<s_PROGRAMME_DESCRIPTION
 element_id="S39" element_name="PROGRAMME_DESCRIPTION">
   <s_PROGRAMME_DESCRIPTION_6
    element_id="A103" element_name="COLOUR_CODE"
    context_id="6">
       <Value>V224</Value>
   </s_PROGRAMME_DESCRIPTION_6>
   <s_PROGRAMME_DESCRIPTION_7
    element_id="A176" element_name="SUBTITLE_FLAG"
    context_id="7">
       <Value>false</Value>
   </s_PROGRAMME_DESCRIPTION_7>
</s_PROGRAMME_DESCRIPTION>
</PMeta>
```

In general broadcast metadata types are highly interoperable due to common standardization efforts and the intention to focus purely on broadcast-related deployment. This is lacking in most multimedia metadata definitions, as they are very generic and are mainly designed for feedback channel network usage. This is not essentially a disadvantage. Due to the fact that broadcasting and feedback channel applications converge, they both complement each other. As an example, DVB-HTML is an adapted version of HTML especially designed for broadcast use. The convergence of metadata definitions used in deployment, consumption and interactions is important when the goal is the successful deployment of interactive TV services.

Example 3.21 (DVB-SI and TV-Anytime). DVB-SI describes how a single MPEG-TS is formed and gives additional information about the stream content. DVB-SI is the basic metadata utilized for the electronic program guide. It contains information about currently running TV programs, follow-up program schedules and describes content of movies and contains other value-added information. DVB-SI is rigidly annotated and is thus based on predefined tables and structures. Extension by broadcaster-defined metadata is difficult and requires granular metadata to obtain more detailed descriptions. TV-Anytime offers this possibility and extends DVB-SI in many terms through its richer descriptors and more in-depth information.

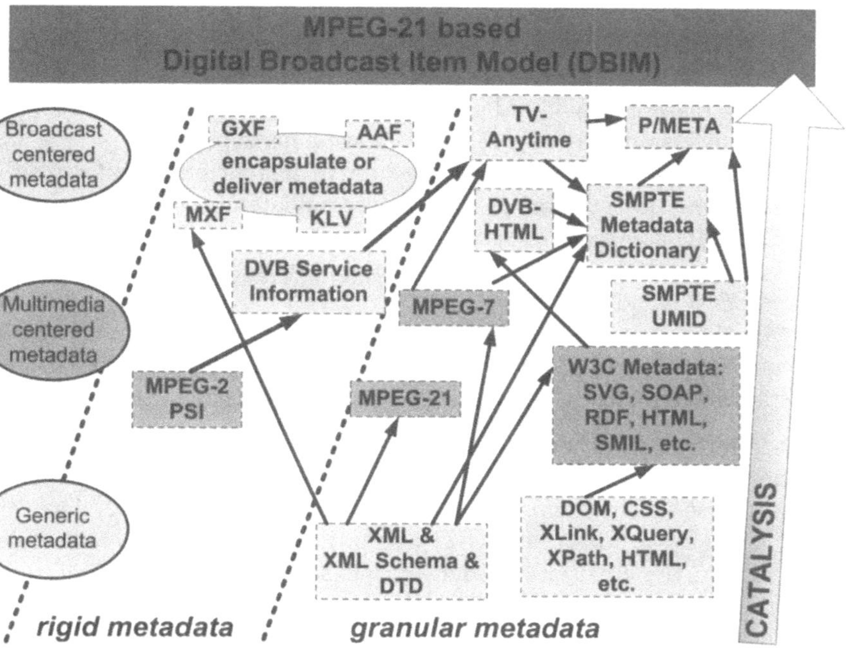

Fig. 3.16. A categorization of the several metadata standards and an illustration of their dependencies

- *system architecture and content management* based on specific broadcast infrastructure or general purpose IT systems including digital rights mechanisms;
- *content representation and packaging models* to unify and abstract multimedia assets and to support life-cycle wide interoperability;
- *transmission protocols* for digital adaptive content packages or their subcomponents for streaming, event-based updates or textual exchange of multimedia assets;
- *service discovery mechanisms* for enabling consumers to find services and creators to announce them;
- *application environment and metadata processing models* to process multimedia assets and to perform application life-cycles in a fully automated deployment chain.

There are many other arguments for integrating and converging different metadata standards. Only one metadata standard provides the capabilities to meet these basic requirements. The application of MPEG-21 has certain major advantages. First, MPEG-21 closes the gap between metadata definitions used in the creation and delivery phases and in the consumption and interaction phases. Second, MPEG-21 acts as a catalyst in combining several metadata standards and provides a single logical digital package throughout the value-chain. In the next chapters we describe the application of MPEG-21 in broadcast multimedia in further detail with the introduction of the digital broadcast item model (DBIM).

Table 3.12. Convergence of metadata structures according to their key strengths

	Metadata Definition and Identification	Lifecycle Phase	System Architecture and Application Environment	Catalysis of Metadata	Transmission and Service Discovery	Resource Encapsulation
AAF	Packaging metadata and management metadata	supports the creation phases	cross-platform and on standard IT equipment	packages and stores metadata but integrates others very loosely	complete file system including simultaneous editing	wrapping of complete packages
MXF	broadcast typical (e.g. subtitles)	supports the creation phases	standard IT equipment and existing broadcast equipment	integrates existing standards (e.g. AAF)	versatile file format	self-containing packages
GXF	simple descriptions of stream content and the transport multiplex	supports the creation phases	none	none	packet-oriented transmission protocol based on FTP	carries complete content packages
MPEG-7	typical multimedia content management descriptors	as multimedia database language or type set	multimedia databases	based on XML schema but restrictive on novel developments	binary streaming	universal multimedia access concept and resource referencing
MPEG-21	digital item methodology	throughout the overall value-chain	own file format and deployable on any XML-based system architecture	encapsulates any type of XML-based metadata	service discovery mechanism, event reporting and binary streaming via MPEG-7 BiM	adaptation mechanism and referencing to resources
W3C multimedia metadata	any type of multimedia presentation	mostly for delivery, consumption and interaction	standard XML deployment architectures focusing on feedback channel networks	application dependent, but the facility to insert generic tags is provided	textual and binary for transmission metadata definitions including RPCs	referencing to content or directly embedded into the multimedia presentation

Table 3.13. Convergence of metadata structures according to their key strengths (continued)

	Metadata Definition and Identification	Lifecycle Phase	System Architecture and Application Environment	Catalysis of Metadata	Transmission and Service Discovery	Resource Encapsulation
P/Meta	syntax and semantic description of attributes	exchange within service providers and third parties	system independent and metadata language independent	integrates TV-Anytime, MXF and SMPTE	binary, textual or streamed, depending on the integrated metadata language	integrates resources dependent on integrated metadata formats
SMPTE	UMID as unique identifier and metadata dictionary	supports the creation processes	broadcast specific equipment	integrates TV-Anytime, MXF, MPEG-7 and MPEG-21 via metadata dictionary entries	indirectly as dependent on the integrated metadata standard and application	indirectly as dependent on the integrated metadata standards
DVB-SI	carrier of service information and MPEG-PSI	delivery and consumption	MPEG encoder including PSI and multiplexer of DVB-SI tables	no catalysis of other metadata standards	required for service discovery and transmitted with MPEG-2 TS	no facilities for encapsulation of resources
DVB-HTML	HTML version for digital TV	delivery, consumption and interaction	HTML features extended by an own application environment	MIME types and XML embedded into HTML documents	similar to the mechanism in HTML	HTML code embedded references
TV-Anytime	broadcast-related metadata for VCR functionality	delivery, consumption and interaction	multiplexed in DVB-compliant streams or over feedback networks	catalysis of MPEG-7	binary, textual or streamed	simple referencing to content
KLV	universal label as identifier	transmission of data	streaming architectures	via key values with very restricted possibilities	packet-oriented transmission protocol	as payload
W3C generic metadata	basis for most granular metadata standards	throughout the life-cycle	standard XML deployment architectures focusing on feedback channel networks	acts as a catalyst for other metadata standards	textual	referencing to content or directly embedded into the multimedia presentation

4

Digital Broadcast Item Model (DBIM)

Recent years have brought tremendous changes in the world of information processing, leading to more sophisticated and automatic models for bringing digital content to the consumer over a hybrid networked environment. There are many new paradigms in the field of broadcasting and multimedia (e.g. digitalization of broadcasting processes, advanced compression technology). Their harmonization, integration and convergence remain a big challenge.

The *digital broadcast item model (DBIM)* is a novel model for managing broadcast content by logical digital packages throughout the broadcast value-chain. The DBIM tries to face the tremendous changes in the world of digital broadcasting. It is based on the ideas from different MPEG-21 standards. The key concept is the extension of the idea of digital item to *digital broadcast item (DBI)*, which is a "configurable, uniquely identified, described by a descriptor language, logical unit for structuring relationships among elements of broadcasted content by referencing to concrete resources of individual broadcast related assets. The end-user perceives it as one entity and as access point to a distributed service pool. "DBIs are DIs, that are especially configured for broadcast use" [128]. It creates a representation for content awareness on various end-user devices. Adaptation of multimedia assets as a key factor for deploying services on low-performance hardware, bandwidth scaling for transmission over IP-based networks, exploration of possibilities in a push/pull environment and metadata definitions are some examples of use-scenarios of the DBI.

The digital broadcast item model (DBIM) is valid throughout the broadcast multimedia life-cycle and its exploration is spread over two chapters of this book. Chapter 4 focuses on general aspects, metadata definitions and behavior models. Chapter 5 emphasizes system architectural aspects in media creation, delivery and consumption.

The key components of the DBIM, valid throughout a typical broadcast system as presented in Fig. 4.1, are the *digital broadcast item model components*:

- *unified DBIM life-cycle and workflow model from creation to interaction* intelligently incorporates existing multimedia management systems and unifies the work-flow;
- *DBI unique identification* throughout the asset life-cycle including globally valid identifiers;
- *DBI metadata building blocks* specifying metadata definitions including different digital item types, metadata structures unique to the DBIM, wrappers for other metadata standards and subsets of existing metadata standards;
- *dynamic DBI process model* for DBIs for describing their behavior and life-cycle;
- *abstract DBIM system architecture including a metadata protocol stack* for seamlessly integrating vertically or horizontally current state-of-the-art broadcast systems for Internet and broadcast deployment including facilities for utilization of standard content distribution standards.

The relevant metadata standards utilized in the broadcast and multimedia industry were presented in the previous sections. Each standard was presented by its defining *organization*, its *standardization process* and the *general purpose*. Even though broadcasting processes rely on standardized procedures, each newly emerging metadata standard has to fit into a complete TV production, from end-to-end. This implies involvement in TV production life-cycle management systems. A deeper view on each metadata format is given by its *collection goal*, *use model*, *utilized technology* and *coding formats*.

D. J. Rayers associates TV production processes with relevant technologies in [151]. O. Morgan introduced an excellent life-cycle model in [134] for media asset workflow within his standardization efforts of the MPEG-7 Systems draft. His description is extended to a more abstract level and convolves the basis of the presented media asset life-cycle model within this chapter. [1] D. Bordwell and K. Thompson's book *Film Art: An Introduction* [25] is an excellent introduction to film aesthetics for an inexperienced reader audience. The art of making films, guiding the reader through different production processes, movie history and fundamental aesthetic aspects are nicely presented. Software development processes, development and implementation methods and different design approaches of technical systems have a key role in the deployment of services in digital television. There are many contributions to this field [129, 93, 55, 130].

4.1 Purpose and Objectives

The purpose of a DBIM is to harmonize several broadcasting related metadata standards to one unified data structure compliant with the theoretical

[1] Andreas Mauthe and Oliver Morgan have used this figure in various presentations and publications.

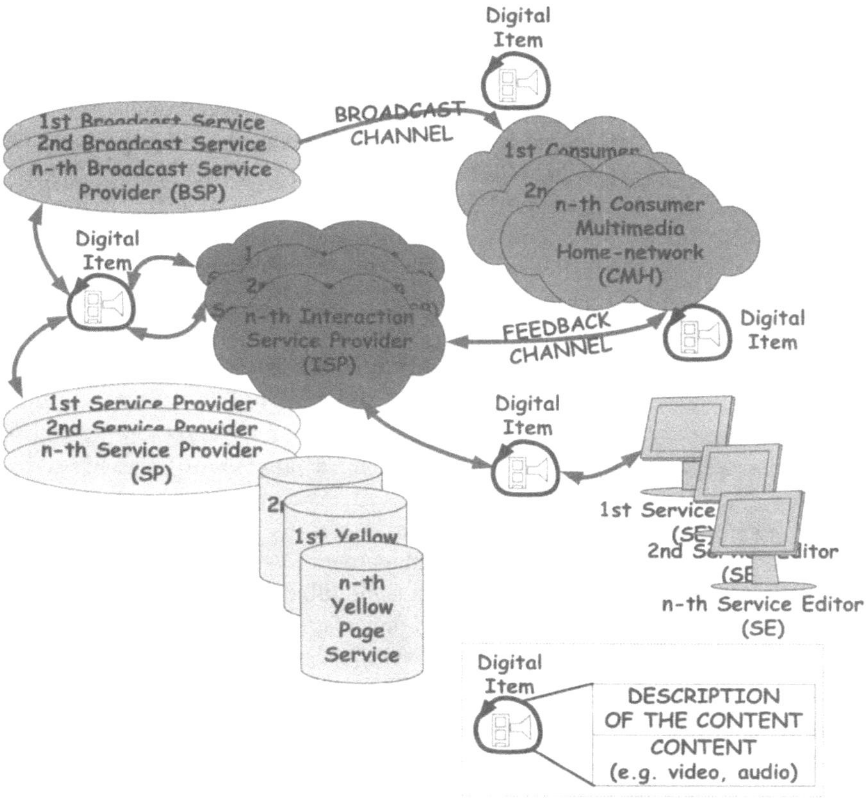

Fig. 4.1. A typical digital television broadcast system

metadata foundations as presented in previous sections. The key features of the DBIM are:

- *harmonization of metadata standards:* Different metadata standards have been developed for use in digital TV systems. Their harmonization and combination to new advanced service types is essential. An MPEG-21 based approach enables the involvement of rigidly defined metatada, generic metadata standards and multimedia content asset referencing through the utilization of the digital item (DI) structure;
- *sharing, development and exchange of broadcast related multimedia assets:* Each consumer will be able to produce content for digital TV. This moves the end-user from a pure passive state to a more active state, enabling him to interact with digital TV content and to produce his own multimedia assets. This requires a general framework, tools and guidelines for how to enable private content productions;

- *provision of an abstract workflow framework:* Representation of workflows of media assets during their whole life-cycle. It provides an abstract framework, where each phase identifies concrete processes. Particular processes of different life-cycle phases are refined into sub-workflows, sub-processes and methodologies, supported by gadgets and tools in performing their predefined tasks. The model provides an easy integration of new business partners joining the value-added service chain. Each phase has its own requirements, characteristics and workflows providing a comprehensive solution for future harmonization and integration of broadcast multimedia assets. The introduction of a metadata-assisted media asset management approach throughout the value-added service chain helps identify requirements for each life-cycle phase;
- *open for integration, interoperability and harmonization of emerging definitions:* Horizontal or vertical integration of new emerging definitions, media asset types, processes, methodologies and gadgets is desired. New media asset definitions and management systems can be easily assigned to each life-cycle phase. It provides an abstract framework for further technical development integration and assigns workflows to each phase. Currently existing media asset management system parts in any form are either integrated as a whole, or with their subparts in the form of processes, definitions or methodologies. Harmonization and convergence of multiple different media asset management systems enable easy integration of systems into existing commercial available repositories;
- *transparent data interchange levels:* Interchange levels and protocols are transparent. There is a clear distinction between media asset interchange levels between partners and machines. Separation of metadata and content flows is unambiguous. This involves enhanced interoperability models for business-to-business, business-to-consumer and consumer-to-consumer transactional services. Distributed services in a distributed service pool, from production to consumer are represented on higher abstraction layers;
- *enabling ontology and taxonomy development:* Categorization of metadata standards according to their strength in the life-cycle model. The development of an ontology is a rather complex task. An abstract life-cycle model will assign easier limitations;
- *rapid content development and distribution:* Development phases of digital TV content become shorter in time.

4.2 Unified Lifecycle and Workflow Model

A general architecture for the deployment of digital TV services was introduced in the previous sections. The DBIM relies on the principal metadata life-cycle model as presented in Chapter 3 in Fig. 3.1. It identified the fundamental broadcast chain contributors and data interchange levels. It is obvious that the first step of concretizing and extending this elementary framework is the definition of a more concrete but still abstract enough behavior description in form of a high-level media asset life-cycle model. Identification of each life-cycle phase leads to a metadata assisted media asset management system throughout the value-added service chain.

The central idea is to describe the sequential phases of development steps through which media assets evolve throughout the value-added service chain. Each phase has its own requirements and characteristics during the production process of an interactive broadcast show in terms of underlying methodologies, processes, workflow management and utilized tools. For example, a real-time play-out of a sports event is tightly bound to time restrictions during the metadata capturing phase. The requirements of distribution of an action movie are more related to packaging and preproduction processes, such as automatic annotation of actor tags or general movie information. Within this section a very abstract lifecycle model for media asset management is presented. Later on, it will lead to the explanation and categorization of multiple metadata taxonomies. Its purpose is to introduce a general purpose media asset lifecycle model valid for media asset deployment in digital television.

It is obvious that digitalization opens more facilities and more contributors, such as multiple BSPs, ISPs, etc., which try to find their place in the value-added chain and to deploy their services with shorter development phases. Rapid service development within a distributed service network requires advanced models of multimedia data management. They enable enhanced interoperability for business-to-business, business-to-consumer and consumer-to-consumer transactions;

A model for this purpose must be general while keeping the long-term objective for reuse in mind. In recent years multiple new metadata standards with different objectives and ability to be integrated vertically or horizontally in current systems have been defined. Each of them provides different benefits, overheads and drawbacks and is optimized for particular use-cases. Each metadata definition identifies other purposes and goals. In a distributed multimedia environment it is desirable to aim at a harmonized and converged multimedia landscape with a high degree of interoperability. Still, lifecycle phases are blurred and not completely deferrable.

But for first system development steps the model provides an abstract enough framework to define basic data interchange interface descriptions between lifecycle phases. This helps to manage future interoperability issues, thus providing interface descriptions for data exchange among different media asset processes, methodologies and tools. The integration of new interchange formats and lifecycle steps is easier. Utilizing an abstract metadata lifecycle model is the first step in identifying the strength of metadata definitions for further categorization, assignment to production steps and unified descriptions of interfaces between each system part. Re-authoring, and more rapid service development is guaranteed by emphasizing this aspect during model development.

Table 4.1. Metadata representation during each lifecycle phase (based on and extended from [151])

Lifecycle Phase	Metadata to Be Captured	Metadata Set Described by	Metadata Coding
Preproduction	commission document	commission data model	XML, requirements document
	scripts, resource plan	text, script data model	document storage system, XML
	plan of program composition (e.g. actors, location)	program plan, data model of a composition	XML, document storage system
production	metadata either captured or generated by the camera	MPEG-7 content management and content description	XML, MPEG-7
	metadata captured by devices other than the camera (e.g. camera movement)	camera movement trajectory (MPEG-7)	XML, MPEG-7
	metadata directly associated at the time of capture	program shooting metadata	MPEG-7
	material segmentation	association of timeline events, multimedia assets and segmented objects	XML, MPEG-7
	metadata ingesting	manually edited metadata structures	MPEG-7, TV-Anytime
Postproduction	review of material (editing and synthesis)	logging data model	XML, MPEG-7 and TV-Anytime
delivery	packaging of multimedia assets	MPEG-21 digital item declaration	MPEG-21 DI, TV-Anytime
	play-out	broadcast play-out configuration metadata	XML, DVB service information, MPEG-7 BiM
	adaptation to play-out facilities	MPEG-21 digital item adaptation, MPEG-21 event reporting	MPEG-21
consumption	resolution and authorization	local metadata asset references, MPEG-21 IPRM	TV-Anytime, URL, XML, MPEG-21
	use and exchange of digital multimedia assets in a multimedia home environment	MPEG-21 digital item adaptation	MPEG-21 SOAP
interaction and transaction	exchange, updating and transmission of interaction services	transactional metadata, MPEG-21 event reporting	SOAP, MPEG-21, MPEG-7, TV-Anytime
multimedia content repository	content management, local resolution, interaction enabler	several metadata descriptors	MPEG-21, XML, MPEG-7, TV-Anytime

Each phase is expressed by a metadata process. Particular processes are executed within each metadata life-cycle phase by utilizing metadata gadget sets to transform, capture, acquire, collect, process, present and disseminate metadata. Currently multiple different metadata standards, dictionaries and definitions exist and introduce a diversity of metadata methodologies into each metadata life-cycle phase. Each of the methodologies makes the whole workflow easier and provides a description and implementation of particular metadata processes. A set of metadata gadgets supports each workflow step by either semi or fully-automated tools for executing tasks on metadata.

4.2.1 Example: Converging TV-Anytime and DBIM Work Flows

Metadata and content flows are strictly separated in the TV-Anytime process model. Four stages of processing are present for creating, publishing, selecting, presenting and handling user interactions. They integrate with the presented lifecycle model into postproduction, delivery and consumption (see Fig. 4.2) phases.

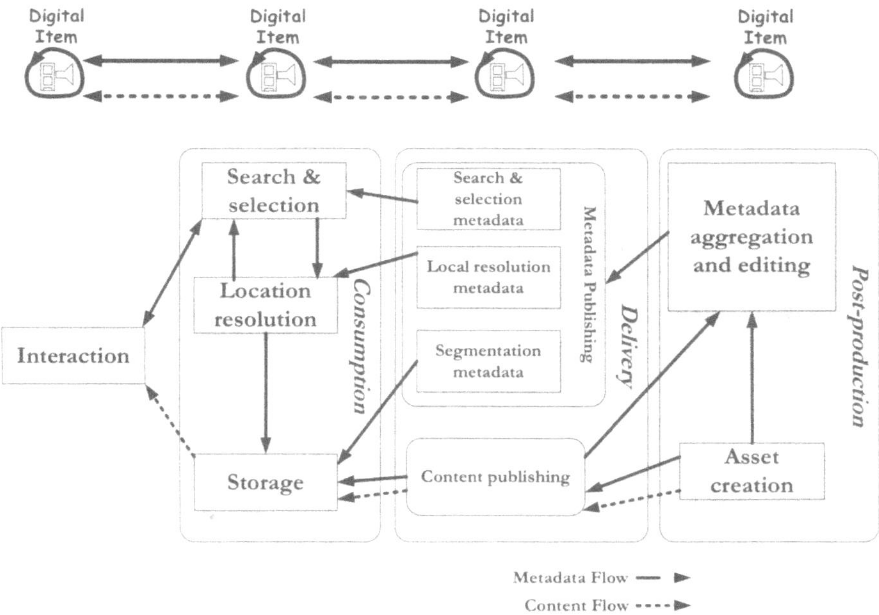

Fig. 4.2. TV-Anytime process model converged with the DBIM workflow (slightly extended TV-Anytime workflow model as described in [81])

4.3 Architectural Components — A More Detailed View

In this section architectural components and elements of the DBIM are explained. Most of the architectural components are part of the DBIM Systems definitions, presented later on in this book. Figure 4.3 gives an overview of architectural components and relates novel approaches to existing solutions.

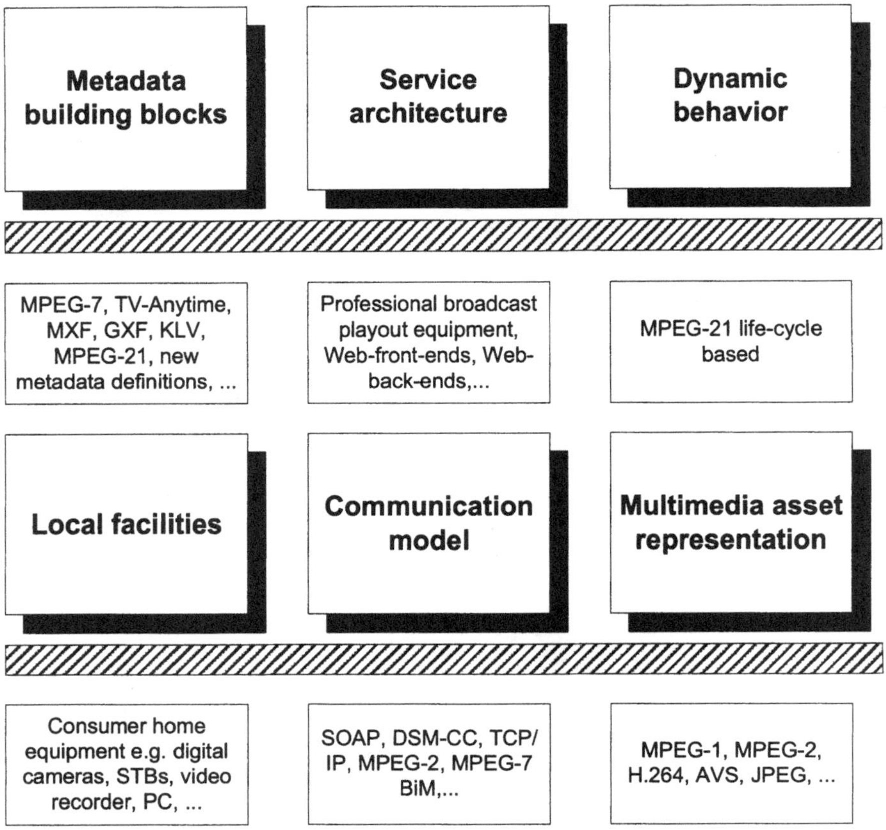

Fig. 4.3. DBIM architectural components related to existing solutions

In principle the DBIM consists of the following elements:

- *DBIM metadata building blocks:* metadata definitions including their categories and data models;
- *service architecture:* architectures for broadcasters, service providers, service editors and interaction service providers to deploy DBIs;
- *dynamic behavior description:* a model that describes the lifecycle in specific application context;

- *local facilities:* equipment and facilities at the consumer side to enjoy services based on the DBIM;
- *communication model:* inter-device communication via a metadata protocol stack including linkage metadata definitions;
- *multimedia asset representation:* a layer model to help to align content representation.

4.3.1 DBIM Metadata Building Blocks

DBIM metadata building blocks are relevant for providing a set of predefined metadata structures utilized for the development of enhanced services. Each building block is applicable for different purposes and is based on existing metadata standards. The central element is a MPEG-21 digital item, packaging a set of relevant metadata definitions and resources to one entity.

DBIM metadata building blocks can be categorized as:

- digital broadcast item types defining item structure;
- descriptive metadata for concrete asset descriptions.

4.3.2 Metadata Protocol Stack — Linkage Metadata Definitions

Different linkage metadata definitions allow reaching a communication goal. A communication goal is arbitrary but often represents the successful exchange of a complete message between parties. Successful sending and reception of an e-mail is an example of achieving one communication goal. They are primarily relevant for configuration management, communication protocol related issues and for reliable delivery. Their utilization is twofold, as they also represent the content and protocol issues of the metadata to be transferred. They also represent concrete items encapsulated in lower-layer protocols as a delivery unit.

4.3.3 Service Architecture

The service architecture includes several elements for professional multimedia asset distribution. Several life-cycle phases are represented within the scope of a logical service architecture (see Fig. 3.1). The components of the service architectures are not bound to a specific party in the life-cycle. This means we introduce only a logical service architecture on an abstract level. Within this chapter we present the framework of a service architecture. The essential components of the service architecture are comprised of a *feedback service architecture*, a *broadcast service architecture* and *local facilities*.

Due to the high abstraction level of the service architecture, as defined within the scope of this chapter, it is important to understand the following:

- the service architecture only abstracts tasks of a distributed architecture;

- each task can be performed location independently by SP, SE, BSP or ISP;
- each architectural component can be, but must not be, uniquely assigned to one partner in the life-cycle;
- sub-components or multiple similar architectures for feedback or broadcast channel can be distributed among different partners (e.g. media creators, distributors, service providers);
- the abstract definition of a service architecture provides a framework for assigning which facilities are required to perform certain tasks;
- separation of tasks, interchange levels and processes are performed through different architectural components.

In the following a brief introduction to the major purposes of the three architectural components is given.

Feedback Service Architecture

The abstract feedback service architecture is utilized mostly by an ISP. But its structural design might also involve other parties such as SEs, BSPs or SPs. The feedback service architecture is relevant for distributing and storing data for the feedback channel connected STB. The feedback service architecture is comparable to a distributed service pool, where value-added content is made available to consumers. It focuses on the provision of services over the feedback channel from the ISP, SP or BSP. On a very abstract level it consists of a multimedia content repository for service storage, service control to control play-out mechanisms and access rights and play-out facilities to interchange services. A multimedia repository stores content and metadata. The service control allows content manipulation and compilation, determines access modes and runs tasks upon multimedia assets. Service access addresses tasks such as access control, billing and accounting, service use monitoring, service play-out server provision (e.g. gaming server) and a service portal front-end maintenance (e.g. HTML server).

Broadcast Service Architecture

The broadcast service architecture provides a broadband multipoint data connection to the consumers. Real-time broadcasting of data to provide multimedia assets over a broadband connection is the task of this service architecture. Its tasks are multiplexing of a high-bit-rate MPEG-2 stream, its distribution, service and application encapsulation, high-bit-rate audio/video/data transmission, etc. Together with the feedback service architecture they can be described as service architecture for the delivery of value-added services to the consumer.

Local Facilities

Local facilities allow access to broadcast and feedback network data deliveries. Due to the increasing amount of various types of multimedia assets that are delivered to the consumer, client architectures have to be implemented in a sophisticated way. Therefore new features, such as computer graphics support, implementation of general HTML browser software, dynamic/static metadata handling support and usability testing of user interface components are a requirement. Authoring tools will support the client development for different purposes. Generic libraries will support different service types multiplexed in an audio, video or data stream. This enables rapid service development and client deployment for advanced service types before transmission.

4.3.4 Metadata Protocol Stack

A metadata protocol stack model enables transparent, reliable and synchronized delivery of metadata structures according to different communication modes. To introduce new formats of delivery, custom protocol suites covering basic requirements in transmitting metadata are required. Their integration into existing specifications and standards helps to avoid the definition of new protocols.

4.4 DBIM Metadata Structures

The DBIM metadata structures are vertically arranged in a multilayer model, which is related to the introduced metadata standards in previous sections. Each layer is self-explaining, has its own atomic unit of processing, provides a set of tools and metadata definitions based on existing or newly introduced metadata standards, describes interfaces to other layers and provides extensibility for new definitions.

Five tools — thus layers — for different purposes are currently defined: *Basic Tools*, *Multimedia Asset Tools*, *Object Tools*, *Service Tools* and *Narrative Tools*. Each tool unifies and abstracts its underlying functionality, metadata model and application scenarios. Each tool consists of building blocks, categorizing sub-metadata families for each tool. Figure 4.4 shows the overall model of the DBIM metadata structures.

Basic entities, definitions and standards are part of the *Basic Tools* layer (e.g. MPEG-7, MPEG-2, image standards, XML). To obtain a self-containing layer, with predefined interfaces to higher layers, a metadata and asset wrapper provide unified building blocks to higher layers in the taxonomy.

Multimedia Asset Tools, building the second layer of the reference model, cover system architectural and physical multimedia asset representation. This might be a multimedia content repository or a movie shot with its spatial and

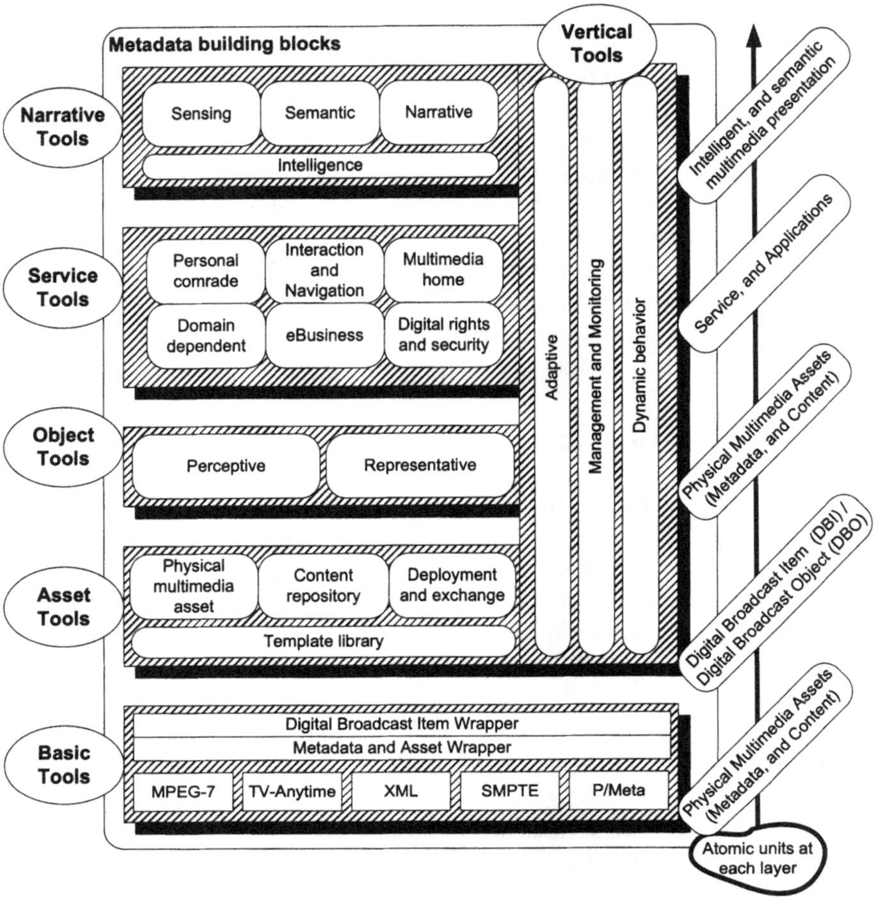

Fig. 4.4. Metadata building block part of the DBIM

temporal aspects. This layer deals with assets as complete multimedia objects (e.g. a complete movie, digital TV broadcast show).

Elements and sub-elements of complete multimedia objects are covered on a higher layer, by the *Object Tools*. Each element of a whole piece of a multimedia object is addressed by this layer (e.g. shots, persons, sound effects) in temporal and spatial alignment to the major contribution line. Add-on services, especially such as applications synchronized with a digital TV broadcast show are a major part of this layer. The purpose of metadata definitions is more related to perceptive and representative purposes.

Service Tools contain metadata definitions for the development of services and applications (e.g. service description).

Structure and semantics are introduced by the highest layer of the reference model i.e. by *Narrative Tools*. They relate different multimedia assets to each other and bring a certain meaning and relevancy into their temporal evolution.

Relevant definitions throughout the multilayer model are *event reporting*, *adaptation* and *management* of metadata structures. Event reporting guarantees continuous updates and the asynchronous/synchronous update of metadata trees on the distributor and consumer sides. The big field of adaptation addresses content adaptation to different local facilities, such as bandwidth requirements, display facilities, input devices, etc. Management metadata, dealing with local resolution, life-cycle management, unified delivery and content identification are also important [123].

4.4.1 Basic Tools

The atomic unit on this layer is a concrete physical multimedia asset, either metadata or content. The goal of this toolset is the definition of a basic toolset of metadata definitions for enabling convergence of metadata on higher levels. Included in this layer are MPEG-7, MPEG-21, TV-Anytime and other metadata definitions as based on XML and XML schema, whose strengths are apparent on higher layers. The functionalities, metadata model and application scenarios are defined in relevant standards, such as e.g. TV-Anytime. Metadata and the asset wrapper provide the abstract DBI structure, a general structure for embedding multimedia assets into a DBI, subset metadata standards in a useful way and interfaces for higher layers.

Table 4.2. Basic tools descriptors

Building Blocks	Example Descriptors
MPEG-7	defined in MPEG-7 standards
MPEG-21	defined in MPEG-21 standards
SMPTE	defined in various SMPTE standards
TV-Anytime	defined by TV-Anytime
P/Meta	defined by EBU
metadata and asset wrapper	types and definitions for wrapping and encapsulating MPEG-7, MPEG-21, SMPTE, TV-Anytime and P/Meta metadata to a digital broadcast item/object

4.4.2 Multimedia Asset Tools

A DBI acts as an atomic unit of processing on this layer for multimedia metadata assets. Multimedia content assets are dealt with as a whole, as a concrete multimedia entity (e.g. whole movies, music tracks, images). They are only dealt with as resources with references to them. Local resolution to actual multimedia assets is relevant for their localization on the basic tool layer, encapsulating lower-layer multimedia assets.

The goal is to define a set of definitions for architectural aspects, content management, content packaging on a higher level, local resolution and involvement of temporal and spatial aspects. Strong incorporation of vertically arranged tools, such as event reporting, adaptation and management is given (e.g. adaptation of multimedia content assets to available network resources, life-cycle management and continuous update through system events). The interface to lower layers is stated through references. Higher-layer interfaces are related to involve perceptive and representative meaning to a DBI.

Table 4.3. Multimedia asset tool descriptors

Building Blocks	Example Descriptors
template library	template descriptions, asset abstractions, dynamic behavior patterns, service patterns, requirement definitions, service templates, representative templates, play-out templates
physical multimedia asset	global identification and referencing, storage definitions, storage patterns, transcoding hints, asset properties
content repository	repository features, database management, storage definitions
deployment and exchange	play-out configuration, interaction descriptors, capabilities, requirements, exchange (e.g. SOAP)

4.4.3 Object Tools

Sub-elements of atomic units on multimedia asset tools are atomic units on this layer. It is important to note that the sub-elements are peculiar. This is due to the characteristics of a layered representation of metadata. It is comparable to the processes of sending data over the Internet. E-mails are encapsulated within numerous IP packets, but the actual IP packet is hidden from the consumer.

This process is similar to a metadata layer model. A broadcast video is tagged with actor names and other relevant object descriptors. Those metadata descriptions are encapsulated in multimedia asset tool metadata, which is invisible at higher layers. The goal and functionality are the sub-sequencing of multimedia content assets by annotating them with higher-layer metadata (e.g. name of the actor, video segmentation, subtitles). The most relevant metadata standard for this purpose is MPEG-7 and its definitions. Interfaces to the lower layer are convolved by invoking perceptive and representative meaning. Perceptive meaning implies the segmentation, extraction of information of existing multimedia content assets and preparing it in a more advanced way for the consumer (e.g. subtitles). Representative meaning reflects the current representation of movie material, as the final result is modified and might require transcoding and conversion to other applicable formats.

Table 4.4. Object tool descriptors

Building Blocks	Example Descriptors
perceptive	visualization features, appearance, asset enhancements, rendering procedures, instantiated VRML or SVG file and related data
representative	metadata containing descriptions/features about visualization elements (e.g. VRML, SVG)

4.4.4 Service Tools

Services built on the lower layers of the metadata model are atomic units on this layer. Services have to be aligned to one broadcast stream, as well as synchronized, and must involve user interactions. This layer structures broadcast, community building, service domain (e.g. e-business services), interaction and navigation and multimedia home metadata to a useful entity to the consumer. Temporal aspects are highly relevant. The interface to the lower layer is built due to the fact that services are built on top of other descriptive elements. An example scenario is a video-overlying display, showing the facility for buying goods currently displayed.

4.4.5 Narrative Tools

Intelligent and sophisticated techniques are applicable on this level to compile a broadcast show personalized for a consumer. Useful narrative pieces built on top of lower layers are atomic units. The definitive goal on this layer is the utilization of general narrative or story-telling patterns to bring meaning into the flow of a presented broadcast show. Methodologies can range from simple intelligence for personalizing content, or more sophisticated solutions for providing semantic/narrative structures. The user experiences a personalized and narrative show suited for her preferences. A sensing block adds features for enhancing the relations of the consumer to the digital content. Content is perceived, rather than being a sequence of bits and bytes on the hard disc.

4.4.6 Vertical Tools

The name vertical tools results from their characteristic of being present on each layer of the general deployment model. Each tool deals with the atomic

Table 4.5. Service tool descriptors

Building Blocks	Example Descriptors
generic service	service identifier, service facilities, creator, management
community	computer-mediated community service descriptors (e.g. chat-room)
interaction and navigation	interaction facilities, navigation path, feedback channel
multimedia home	home service descriptors, home space abstraction, consumer end-devices, equipment interoperability, home multimedia center, leisure content descriptors
personal comrade	personal profiles, service adaptation to consumer behavior, consumer behavior descriptors, personalization
digital rights and security	protection scheme, rights management, license
eBusiness	payment modalities, information exchange to business partners
domain dependent	various descriptors related to specific service types

Table 4.6. Narrative tool descriptors

Building Blocks	Example Descriptors
sensing	digital content enhancements, ambient definitions, natural interface descriptors
semantic	structure, grouping of assets, annotation of meaning to assets
narrative	story patters, piece matching patterns, story evolvement, narrative mosaics
intelligence	profiles, intelligent patterns, algorithm descriptors, data mining definitions

unit present on the applicable layer. Adaptation is relevant for manipulating atomic units to present them scaled to the available resources (covered by the MPEG-7 universal multimedia access (UMA) definitions and the MPEG-21 DI adaptation mechanisms). Management deals with the control and monitoring of atomic units through database entries, log-files, etc. Event reporting introduces metrics and measurements for describing happenings in a digital TV deployment environment (e.g. number of consumers, system faults and channel performance).

Table 4.7. Vertical tool descriptors

Building Blocks	Example Descriptors
adaptive	adaptation mechanisms (mostly based on MPEG-21)
management and monitoring	usage monitoring, content descriptors throughout the life-cycle
dynamic behavior	update descriptors, dynamic behavior patterns, temporal–spatial behavior descriptors

4.5 Digital Broadcast Item (DBI)

An MPEG-21 based DBI is a configurable, uniquely identified logical unit described by a descriptor language for structuring relationships among elements of broadcast content. It includes references to concrete resources of individual broadcast related assets. The consumer perceives it as one entity and as an access point to a distributed service pool. DBIs are DIs especially configured for broadcast use [128].

Figure 4.5 shows an example of an instantiated *digital broadcast object (DBO)* as based on a DBI. Each DBI is structured into a container part, acting as a placeholder for multiple subitems. Subitems and items consist of a description part and a resource part. The description part is based on metadata definitions as presented in this section. The resource part refers to a multimedia asset, thus a physical resource.

Very elementary types of the DBIM are as follows:

- A *digital broadcast creation item type (DBCIT)* finds its application during preproduction, production and postproduction. It is the atomic unit of exchange during these lifecycle phases and packages assets during creation. This item type relates to the exchange of assets between partners taking part in the creation phase of broadcast content.
- The *digital broadcast multimedia home item type (DBCMHIT)* describes multimedia assets consumed, processed, created and stored at the consumer side. It is specifically designed for the digital consumer home. It is directly processed at the consumer terminal and adapted to the specific needs of various platforms.
- A *digital broadcast feedback item type (DBFIT)* relates to feedback channel related deployment descriptors by referencing or describing interaction-based services. It is the atomic unit of exchange within feedback channel communication and relates to consumer-to-business and consumer-to-consumer exchange of data.
- A *digital broadcast TV item type (DBTVIT)* packages content during the delivery of services to the consumer. Its application area is the deployment of digital TV content and the catalysis of play-out information in a real-time broadcast environment. It is the atomic unit of exchange between the broadcaster and the consumer within a broadcast channel network. If

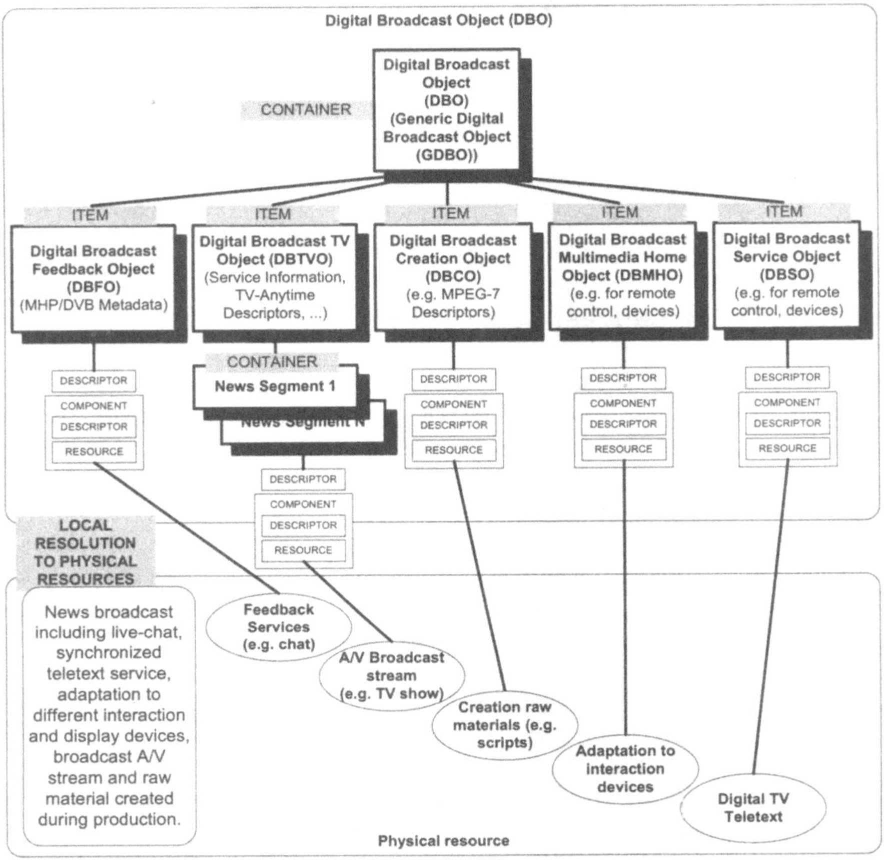

Fig. 4.5. Example for an instantiated digital broadcast item (DBI)

a broadcast show is deployed on feedback channel architectures, its task is to package broadcast content for this deployment mode.

- For created services, a *digital broadcast service item type (DBSIT)* provides data structures and metadata definitions for specific services. This includes global identifiers and mechanisms to group various services.
- The *digital broadcast generic item type (DBGIT)* provides a minimal descriptor set valid throughout the broadcast value-chain. This type is purely based on the MPEG-21 digital item declaration with minor extensions for identification and broadcast multimedia asset descriptors, among others. This item type is minimal and its validity is guaranteed throughout the overall broadcast life-cycle.

4.6 Dynamic DBI Process Model

Within the previous sections several aspects of a DBIM were introduced. It is essential to understand its behavior throughout the life-cycle. Different item types have different purposes and are adapted to the needs of different actors during the broadcast life-cycle. Figure 4.6 shows the overall life-cycle of an instantiated DBO.

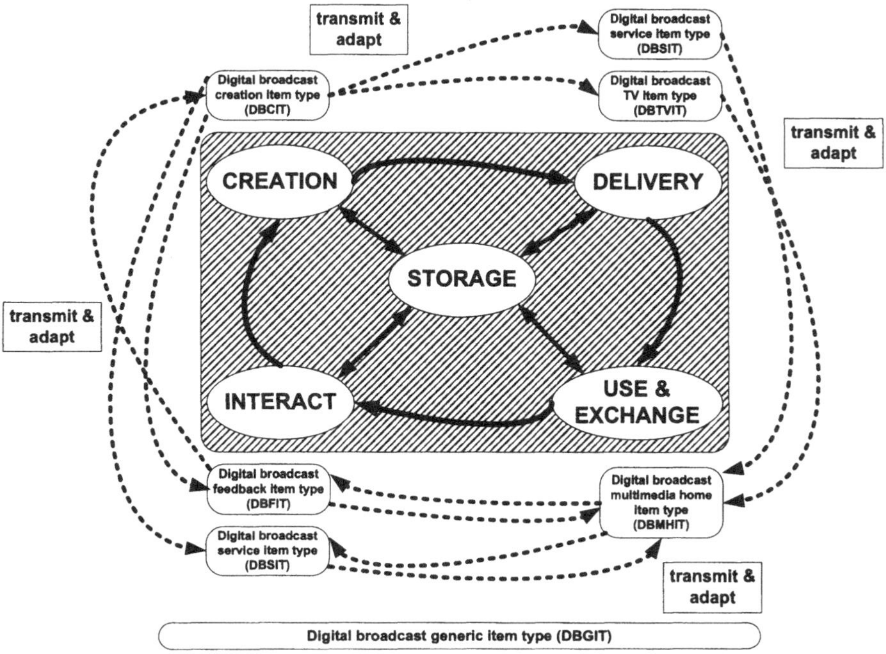

Fig. 4.6. Applying the DBIM throughout the value-chain

All the information contained by the different item types is not required throughout the overall value-chain. For example a consumer does not need the raw scripts or discarded video parts. For a creator they might be still relevant during the creation of a new broadcast show. Only the relevant metadata definitions are guaranteed to be visible during the actual life-cycle phase by limiting item liability and introducing digital broadcast item phases.

4.6.1 Different Item Types in the Metadata Lifecycle

Each item type is only valid throughout one lifecycle phase. Bringing it to the next phase means to transmit, parse and adapt it to the specific needs within this lifecycle phase. This means extracting the essential data and adapting them to the type required during the new phase. Each DBI life begins with instantiating a *digital broadcast creation item (DBCI)* to a *digital broadcast creation object (DBCO)*. It is based on the digital broadcast creation item type and contains a minimal set of descriptions required during preproduction, production and postproduction. The DBCO acts as a container for arbitrary components added during the creational phases. The instantiation and creation of a *digital broadcast service object (DBSO)* is essential. Several created DBSOs represent a bundle of services embedded into a DBCO.

Example 4.1 (digital broadcast creation object). A broadcast show consists of one A/V stream, associated teletext pages and two applications that are running in parallel to the A/V stream. Three DBSOs are present: one for the A/V stream and two for the applications. All DBSOs are embedded within one DBCO acting as a container for the complete TV show.

During play-out two DBI types are essential: a DBCO and its subitems are transformed to a *digital broadcast TV object (DBTVO)* containing all information required to deliver a compiled broadcast show to the consumer and an optional *digital broadcast feedback object (DBFO)* is deployed for the delivery of Internet services. The DBFO initiates the lifecycle for feedback services.

Within the consumer domain another object is relevant: *digital broadcast multimedia home object (DBMHO)* is deployed within the range of a consumer multimedia home network. It contains services visible to the consumer and acts as the service space with which the consumer is interacting. Communication between the DBFO and the DBCMO is not very relevant, as several feedback and interaction architectures enabled via an Internet connection are used at this stage.

It is important that items communicate with each other during their lifecycle. Communication is arbitrary but results in the construction and reconstruction of whole metadata trees or their subtrees. This enables e.g. the digital broadcast multimedia home object to communicate with a digital broadcast feedback object and to update and exchange information. An application scenario is the exchange of licensing information between the consumer and a rights clearance authority for music files.

4.6.2 DBO Phases

A DBO performs an own "item" lifecycle at each phase in the metadata lifecycle model. In principle it starts with its creation and the DBO life ends with its killing. Different item types are applicable at each step and perform a common abstract process. Figure 4.7 shows the state diagram of a DBO process performed in each lifecycle phase.

The following states are present:

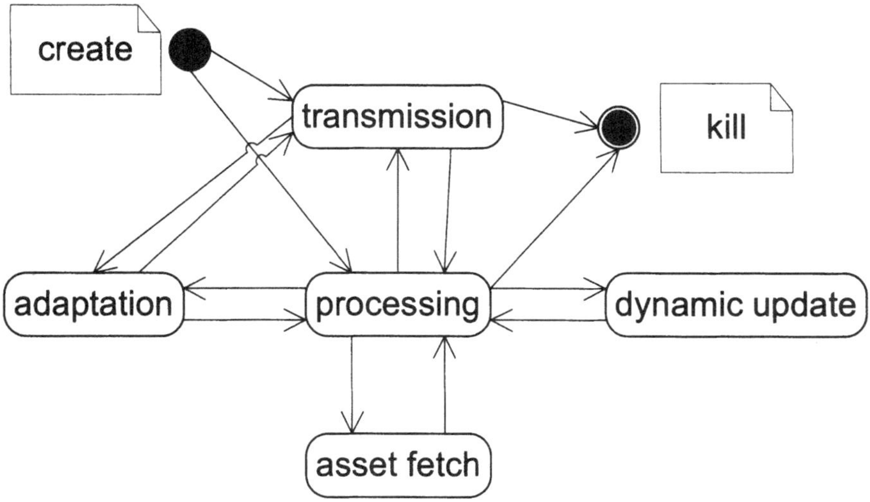

Fig. 4.7. State diagram of a digital broadcast object in each lifecycle phase

- *create:* creation of an instantiated digital broadcast item and its related data structures;
- *transmission:* obtaining or fetching a DBO either per a push or push/pull scheme including parsing processes (e.g. transmission of a service encapsulated in a DBO to the consumer device);
- *adaptation:* adaptation of the DBO after fetching it (e.g. adaptation of the DBO and its encapsulated service to the consumer device capabilities);
- *processing:* performing the actual application or service lifecycle (e.g. executing the service encapsulated within the DBO);
- *asset fetch:* local resolution and referencing to assets packaged by the DBO (e.g. initializing the streaming of a video stream as referenced via a DBO);
- *dynamic update:* updating and event processing for the DBO (e.g. parsing of events sent over the feedback channel and updating the DBO);
- *kill:* terminating the lifecycle of a DBO and freeing the resources required to execute the DBO and its related services.

5

Metadata System View

As previously described, the linkage tier connects nodes in a distributed networked environment. Metadata transmission, protocol stack implementations and deployment architectures are major concerns within this chapter. The question is how to transmit and distribute metadata in arbitrary forms. Previous sections developed certain metadata taxonomy and introduced different metadata types. Within the scope of this chapter we discuss metadata utilized for deployment configuration, streaming and encapsulation in multimedia asset delivery.

This chapter focuses on the following points:

- development of a generic XML metadata architecture for digital TV for the deployment of the DBIM;
- introduction of relevant communication modes relating to synchronization, object serialization, binary representation, etc.;
- presentation of a metadata protocol stack model for the distribution of metadata, especially focusing on application layer protocols (e.g. SOAP, MPEG-7 BiM);
- definition of an abstract service architecture for feedback channel enabled digital TV equipment;
- abstract architecture for broadcast channel architectures;
- abstract lightweight client implementation for processing metadata;
- consumer multimedia home platform.

Each node in a distributed environment is interconnected over well-known protocol stack architectures. The linkage layer provides linkage models, such as synchronization, transmission techniques, etc. Pieces of information require linkage models to fit them into contemporary structures in a networked environment. Simple examples are links on the World Wide Web, connecting Web pages under one structure: the Internet. Each protocol is designed as a multilayer model. Metadata protocol suites enable optimized transmission of metadata either over the feedback or broadcast channel network. Some protocol types are hybrid and enable transmission over both channel types.

Much related research is done in defining system architectures. On an abstract layer, research works can be categorized into *messaging protocols* defining data structures for transmission, *architectural developments* for processing data communication and *communication mechanisms* required to maintain communication (e.g. synchronization).

Three protocol types are relevant for deploying and interchanging messages:

- *Simple Object Access Protocol (SOAP)* [160],
- *Web Services Description Language (WSDL)* [170] and
- *Binary Format for MPEG-7 Description Schemes (BiM)* [4].

SOAP is a lightweight XML-based protocol for the exchange of Unicode text format messages encapsulated in FTP, SMTP, POP and HTTP in a decentralized, distributed networked environment. It is defined in [160] by the W3C. Still, the description of message structures requires maintaining the service goals. Thus, guidelines and rules for successful communication between nodes to achieve service goals is an ultimate requirement. XML-based WDSL files, as standardized by the W3C in [170], abstractly describe interfaces of service providers and their communication rules for successfully achieving a service goal. A yellow page directory service (UDDI) enables searching for service providers as well as automatic searching for services. It acts as a certain kind of online "service phone book".

For further reading we refer to [102, 172, 170, 60]. A binary format for the transmission of metadata content is provided by MPEG's BiM for dynamic updating of metadata data trees as defined in [4]. General internetworking is extensively described in [34]. This excellent work provides a good source for very deep understanding of internetworking.

General metadata-driven system architectures are described in [17]. Feedback architectures and Web service deployment research have been extensively covered previously. Excellent descriptions can be found in [159], covering multimedia streaming servers and their implementation. For Java-based feedback architecture deployment, especially for JSP-based environments, we refer to [61, 62], which provide an excellent summary. Broadcast related service architecture and digital TV client implementations are described in [68, 165]. Communication modes and real-time software development are described in H. Gomaa's book "Designing Concurrent Distributed, and Real-Time Applications with UML" [93]. E. Bertino and F. Ferrari provide an interesting collection of synchronization mechanisms in [24].

5.1 Characteristics of the Linkage Tier

The linkage tier has several characteristics:

- each entity within the distributed service pool is represented by nodes. Interconnection of nodes can be either equal, such as in peer-to-peer networks, or based on client–server schemes;
- from the functionality point of view each node has capabilities to process metadata and depending on the performance capabilities, a lightweight or heavyweight implementation of a metadata protocol stack;
- clear separation between logical and physical model guarantees location independence and platform independence;
- due to the distributed characteristics, where multiple devices with different operating systems are interconnected, platform independent protocol and architectural design are inevitable. Adaptation to network bandwidth and destination platforms is a major concern;
- different communication modes provide synchronization, streaming and communication models for converging multimedia assets to one entity perceived by the consumer;
- textual and binary incremental or complete updates are enabled depending on the utilized protocol type;
- versioning to keep track of changes and update mechanisms;
- digital items are the basic entity to be interchanged between entities. They hold the information required to adapt multimedia assets to different platforms, hold authentication and access rights schemes and provide adequate versioning models.

5.2 Metadata-Based Service Architecture

The service architecture is relevant for providing basic infrastructure to deliver, store, distribute and verify service deployment. Each partner present in the value-added service chain has its own requirements and needs for deploying his services. Whereas the BSP addresses more questions concerning postproduction, the eTailer has to be able to perform rights clearing and access to service types. The final setup covers several needs of each partner into a unified architecture, where most communication is based on an application layer protocol. The digital items are used as the unit of exchange. This is enabled by providing an enhanced feedback and broadcast architecture capable of automatic or semi-automatic handling of digital items. Each partner relies only on digital exchange standards that are encapsulated in the MPEG-21 digital item declaration.

The realization of such service architecture focuses on installation, testing, configuration management, server farms, database structures, performance analysis, migration of existing repositories, etc. The underlying hardware related questions and the software architecture are handled during their development. For service testing purposes, technical equipment has to provide facilities to configure, install and deploy services to enable developers to test, verify, deploy and realize their services in a test environment.

A sophisticated platform solution enables complete metadata-based communication between each partner in the value-chain, over arbitrary protocol suites. Real-time deployment, where the whole life-cycle is considered within the distributed service network demands utilizing the feedback channel, the broadcast channel and local hardware facilities at the consumer side. For the development of distributed services it is essential to introduce synchronization mechanisms for services distributed among the consumer, the SP and the BSP. Therefore it is essential to introduce synchronization models for distributed services.

Delivered broadcast services are presented in a technology hiding and usability-friendly manner to the consumer. Feedback channel invocation is based on different interaction modes. It enables transparent, synchronized and reliable interaction between the different partners in the value-chain.

5.2.1 Logical Feedback Channel Architecture

Distributed logical feedback channel architecture (see Fig. 5.1) requires contemporary and partly standardized ways for providing access to multimedia assets over general interchange protocol types. Most standard Web-deployment architectures are used with enhanced capabilities for dealing with metadata structures. Thus the basic architecture consists of a Web-server, dynamic HTML capabilities, XML content servicing, database systems, etc.

Unlike typical deployment environments, metadata-based feedback architecture requires additional facilities. They include special facilities, such as a DI-based business logic, enhanced content manipulation facilities (e.g. for metadata extraction and real-time streaming), architecture back-end access control, etc. The key element is the multimedia content repository for storing multimedia assets and additional consumer data in a database system. Native RDBS database systems do not provide adequate solutions for metadata storage. Special metadata repositories optimize the access, storage and management of metadata structures.

Physically, the logical feedback channel architecture must not be located at the service provider side. Small-scaled versions can also be part of multimedia home equipment, or even be implemented in digital TV equipment. Their purpose and requirements differ from the commercial point of view. They only support basic functionality for managing personal multimedia assets (e.g. holiday image database, video collection).

The feedback architecture touches more interaction facilities, but can also be utilized to control broadcast content play-out. In the general lifecycle model it covers mostly multimedia asset delivery, postproduction and raw-data repositories. In general, the logical feedback channel architecture can be divided into four key levels of realization with different requirements and characteristics. Their basic functionality and requirements are stated in the following sections.

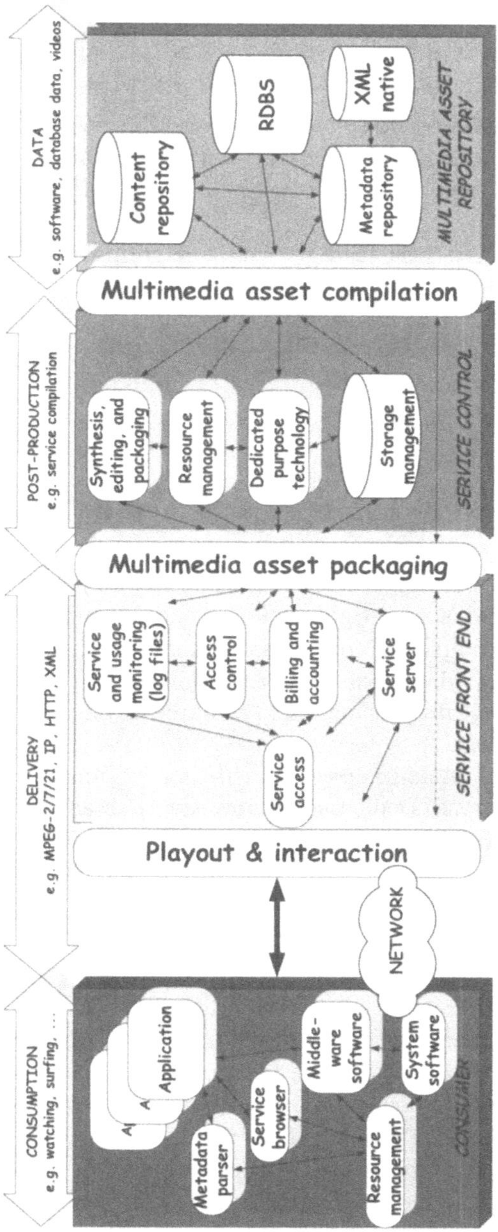

Fig. 5.1. Logical feedback architecture [127]

Service Front-End

The first access point for the consumer to value-added services, either for feedback purposes or to obtain new services, is the service front-end. In principle, it is a typical Web-portal that performs tasks such as usage monitoring, access control management, billing and accounting and handling of special service servers (e.g. streaming solutions for Video-on-Demand services). Requests are forwarded to the back-end structures of the overall service architecture. In an implemented environment mostly HTTP have to be handled. The user interface can be simple Web pages, WAP pages or other XML related visualization facilities (e.g. SVG).

The most advanced task in digital item handling is to obtain and send them over transmission protocols capable of carrying digital items. This implies facilities for adapting multimedia assets into a format the consumer's device can handle. As this level acts as a front-end, it acts more as a delivery layer, rather than as a configuration tool. The actual configuration is done by resource management facilities within the service control layer. Still, its play-out and monitoring are done within the service access layer.

Service Control

In general, the service control level catalyzes user requests after the service front-end has cleared access rights, billing and accounting issues. Content packaging and the compilation of multimedia material is its entire task. This involves preparation, adaptation and protection of multimedia assets, as well as dealing with storage management and resource management of the feedback architecture in general.

New emerging technologies provide real-time streaming content modifications and manipulations. Dedicated technology, such as metadata extraction facilities and narrative script compilers, provides sophisticated solutions for performing these demanding tasks. The compilation and packaging of DIs is done on this layer.

Multimedia Asset Repository

The purpose of the back-end of the architecture is to provide storage facilities for different types of multimedia assets; metadata, data and content. Its requirements range from low-profile solutions up to high performance data warehouses. It is a passive element, thus media compilation is done on other levels of the feedback architecture. It therefore acts in response to requests from other levels of the feedback architecture. Query services are the major communication mechanism with this layer. Legacy data are databases is fitted in as well as other related data structures.

5.2.2 Logical Broadcast Channel Architecture

The logical broadcast channel architecture (see Fig. 5.2) for a metadata-based play-out covers most lifecycle phases of the general metadata lifecycle model (preproduction, postproduction, production, multimedia asset repository and control and monitoring). Each phase is covered by different parts of the general architecture. Figure 5.3 shows three video channels, uniquely identified by their PIDs or DVB service locators. The DSM-CC mechanism delivers metadata structures together with the services.

Front-End

In a DVB-compliant system design, the front-end accumulates the DVB core system for terrestrial (DVB-T), cable (DVB-C) and satellite (DVB-S) digital TV. The input of this architectural building block is graphics, audio, text or video, that is delivered via Internet or a DVB-compliant output configuration. Independent of the delivery system, MPEG-2 TS is the carrier for digital TV multimedia assets. Containers structure MPEG-2 video, audio and data services to more flexible combinations. Their construction is described by DVB-SI and error protection is guaranteed by a common *Reed–Solomon (RS) forward error correction (FEC)* system. The central element in a real-world play-out system is the *(re-)multiplexer*.

Besides the creation of a DVB-compliant MPEG-2 output stream, its feature is to involve scrambling, *conditional access (CA)*, bandwidth optimization and the creation or insertion of DVB-SI and EPGs. MPEG-2 TS or ASI streams usually act as inputs. Available broadband Internet connections enable Internet play-out of digital TV streams. Transcoding, converting high-bit-rate streams to lower bandwidth consuming formats (e.g. transcoding MPEG-2 to RTP-based MPEG-1), results in lower bandwidth consumption by decreasing video/audio quality.

The central element to perform and initiate transcoding processes is the *multimedia asset adaptation (MAA)* engine. The MAA's intelligence is to acquire data about bandwidth availability, play-out facilities or service requirements that enable sophisticated initialization of transcoding or adaptation processes. Multiplexing data services — applications, services and files — is a responsibility of the DSM-CC carousel generator. Usually backup mechanisms support the frictionless output of a continuous data stream, which can be multiplexed into the final play-out MPEG-2 TS. Control and monitoring is an essential instrument to guarantee frictionless and trouble-free multimedia asset delivery. Feedback service architectures support the broadcaster in retrieving Internet data interactions, as well as automatic system integration throughout the lifecycle and accessibility for third-party providers.

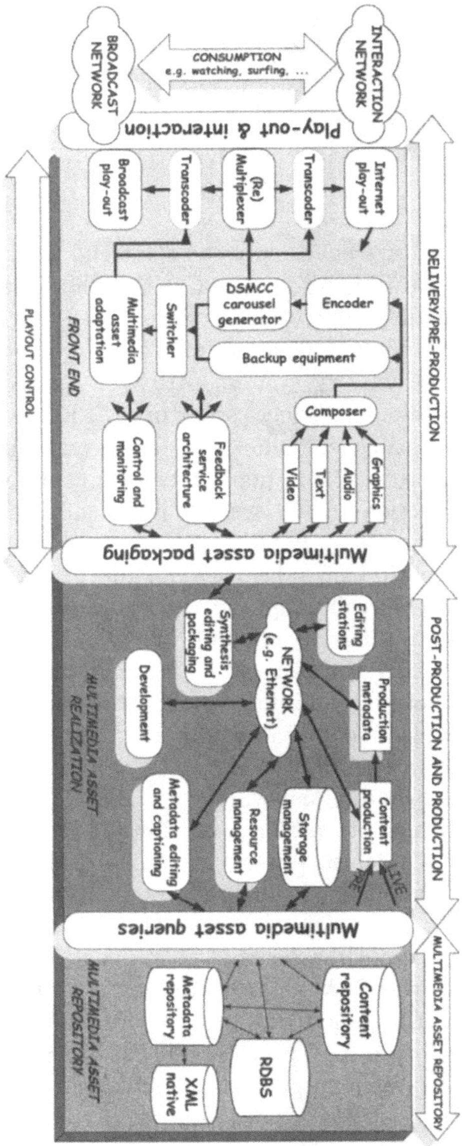

Fig. 5.2. Logical broadcast architecture

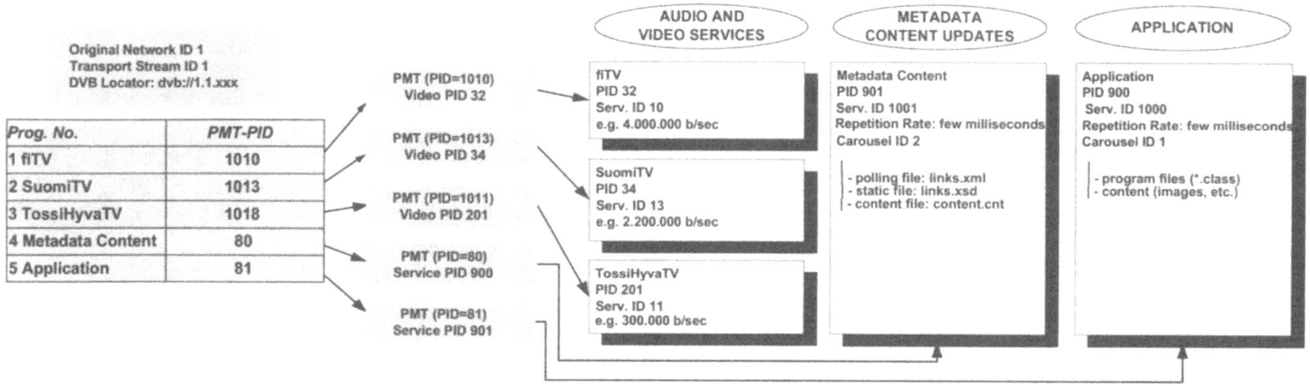

Fig. 5.3. Sample play-out configuration, consisting of three video streams and two data services (one utilizes metadata transmission modes and the other transmits applications)

Multimedia Asset Realization

The creation of broadcast content, its processes and the compilation of multimedia assets in artistic means is performed on this level of the broadcast service architecture. Multimedia asset sources are either live or pre-stored multimedia content. Important for metadata architectures is the synthesis, editing and packaging stage, which encapsulates content and metadata structures for reuse and delivery preparation.

Metadata sources are either production metadata, advanced metadata editing or captioning stations. Their function is the semi-automatic extraction of metadata that has to be compiled into an intelligent multimedia presentation. The rest of the level is mostly equivalent to the logical feedback service architecture.

Multimedia Asset Repository

The back-end of the broadcast service architecture is built similarly to the feedback architecture. Tougher requirements (e.g. access delays, hardware resources, management) and limitations guarantee problem-free editing and delivery processes. An example configuration is shown in Fig. 5.4.

5.3 Metadata Protocol Stack

The development of an XML-based protocol stack implies encapsulation facilities for packaging textual and binary XML data. The following requirements have to be met for the implementation of an XML-based protocol:

- application and presentation layer transmission protocol,
- binarization and serialization facilities of multimedia assets,
- streaming of metadata,
- support for textual and binarized metadata formats,
- event-based update mechanism and
- validation and error recovery

N-tier service architectures are currently state-of-the-art in Web service deployment architectures. They adopt their design principles from component-based software development concepts, where each component is implemented for a specialized context. This compares to distributed services, where each service logically performs another business task.

Text-based communication between different entities in a distributed service pool is based on HTTP/HTTPS [71]. The usage of HTTP(S) as a message transportation protocol is easily explained, due to its easy usage, inexpensiveness, easy encryption and encapsulation of other text or binary data types (e.g. metadata in form of XML, images).

In this context SOAP can be referred to as XML over HTTP, by carrying textual metadata elements between different nodes in a networked environment. To deploy and identify communication goals of nodes with SOAP capability the *Web Service Definition Language (WSDL)* enables node identification and handshake definition to obey the final goal of a communication session. A more complex, but sophisticated solution to interchange metadata in binary format between nodes is the *Binary Format for MPEG-7 Description Schemes (BiM)* which enables dynamic updating and incremental transmission of metadata data trees.

5.3.1 Abstract Metadata Protocol Stack Model

The *International Organization for Standardization (ISO)* has standardized a generic model as the *Reference Model of Open System Interconnection*, commonly referred to as the ISO/OSI model. Its idea is to layer protocols, where each layer has a different functionality. Similar to the introduced tier model of layering metadata purposes, it is defined to describe the purposes of each protocol, rather than its implementation. It is a model for how to categorize and integrate protocols into the existing ones.

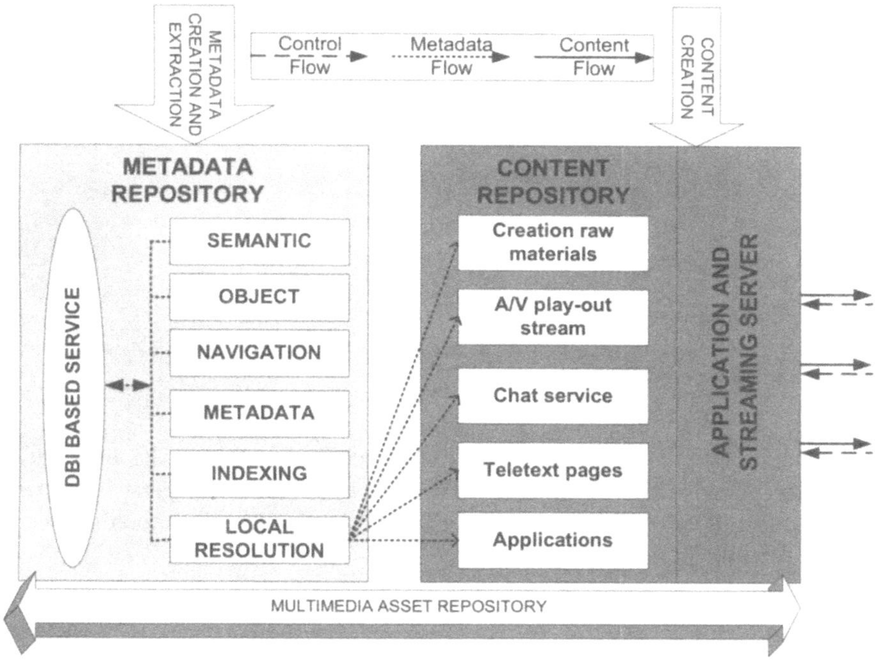

Fig. 5.4. Multimedia repository example for a DBI as presented in Fig. 4.5 (very freely after [97, 119])

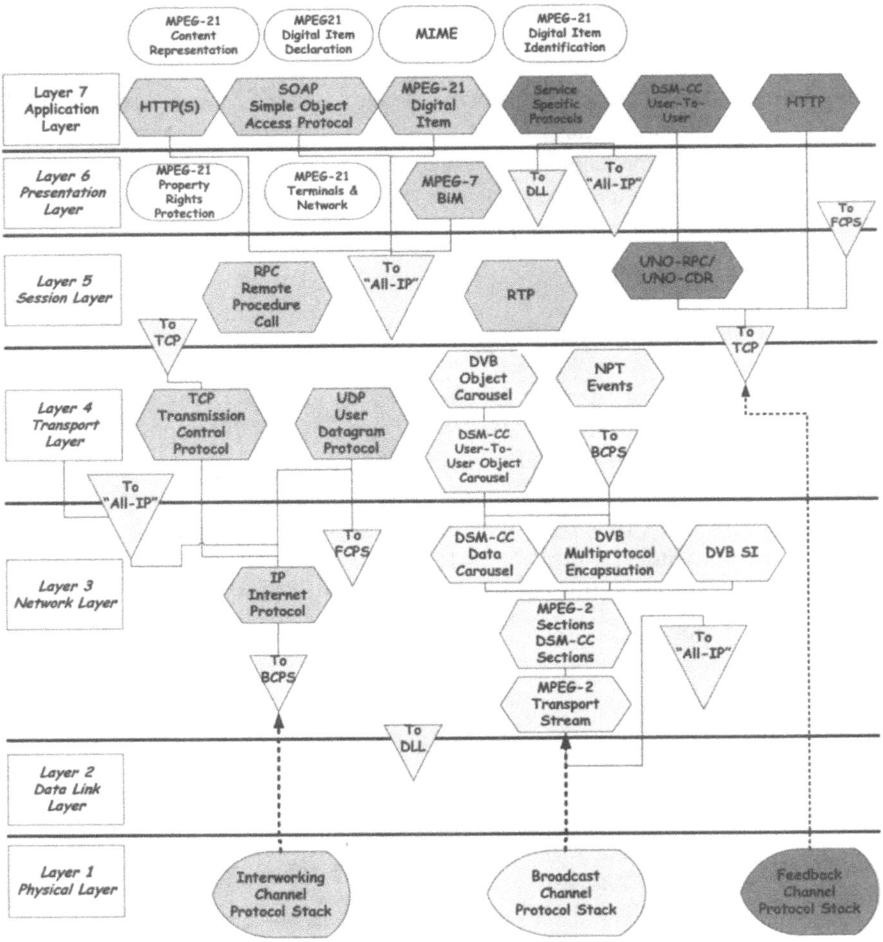

Fig. 5.5. ISO/OSI reference model of open system interconnection enriched by typical broadcasting protocols

Most of the metadata deployment protocols belong to the application, presentation or session layer protocols. Figure 5.5 shows the generic reference model and aligns metadata transmission modes. Internet protocol suites are also shown in the figure.

5.3.2 Internet Protocol Suite

Current developments converge several protocols over an "All-IP" layer where IP encapsulates multiple other protocol types as payload. IP is very restricted in its use and introduces basic networking principles, such as packaging of other protocols in its payload, routing and address resolution. Today another protocol emerges as a container for service related communication protocols.

The "All-HTTP" layer seems to have a similar development as IP by acting as a container of textual format-based communication models on the application layer. The Internet protocol suite provides facilities for transmitting either push or pull content types. Several protocols act as carrier for any type of content. Example protocols belonging to this category are IP, UDP, POP, SMTP, RTP, RTSP and RSVP [148, 147, 139, 16, 158, 29].

5.3.3 Transmitting Metadata over Broadcast Channel Protocols

Transmitting a DBI within a high-bit-rate MPEG-2 TS might facilitate these different transport mechanisms (from [125] and based on [68]).

1. *DSM-CC Object Carousel:* The DSM-CC object carousel is used for transmitting metadata files as a whole or in pieces to consumer device clients. The broadcaster can utilize one of more object carousel(s), uniquely identified by a DVB locator (e.g. `dvb://44.3.39/carouselname`) and a carousel ID/carousel-name. Metadata descriptors are constantly multiplexed. This also includes consistent versioning of metadata files. A constant polling at the client side is needed for performing checks for version changes. Currently two modes for transmitting metadata to the client are possible: textual or binary XML files as MPEG-7 compliant BiM structures. Soft-deadline restricted mechanisms apply at the broadcaster side for updating relevant data structures. An event mechanism is responsible for transmitting version changes to the client. This technology enables a maximal rate of shot-accurate updates. This very powerful and sophisticated mechanism provides updates to the client, but restricts it to simple updating. Constant streaming of metadata is not enabled. But its easy reliability is a great advantage.
2. *Private Section Mechanism:* Encapsulation of binary metadata into private sections within a MPEG-2 TS is a very interesting solution for streaming metadata to the consumer device. Updating, streaming and facilitating advanced hardware architectures are enabled by this mechanism. To fulfill hard-deadline constrained real-time transmission systems, longer client setup times, more complex client architectures and higher computational performance are required.
3. *NPT Events:* Each NPT event is uniquely identified via an event ID (e.g. `dvb://4.4.3;32`) and bound to a concrete service. NPT events provide data encapsulation mechanisms of either raw or binary metadata. A very simple and reliable solution for metadata delivery is enabled. Especially the signaling type of usage, enabling advanced scheduling and synchronization models is a specific feature of NPT events. But NPT events provide more: time-base simulated event monitoring, scheduling of metadata

announcements, time-base synchronization and carriage of metadata time-base synchronization descriptions.

Figure 5.6 shows an overview for a complete teletext service delivered to the consumer.

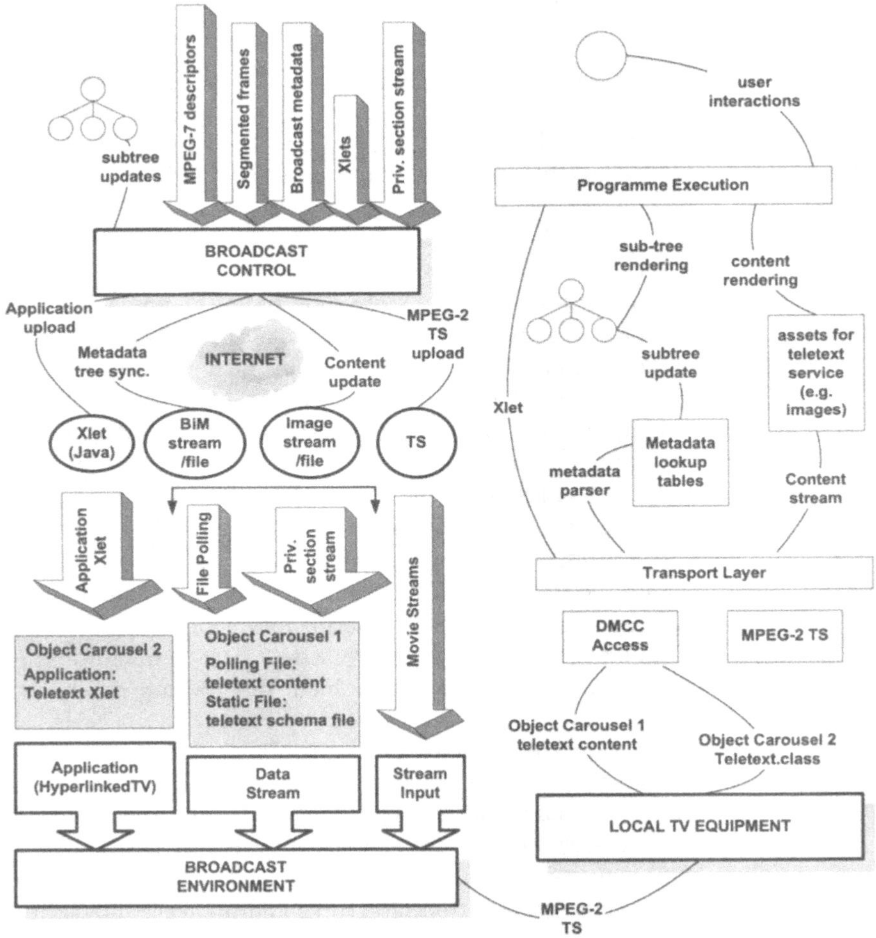

Fig. 5.6. Transmitting a complete teletext service including metadata definitions and content assets to the consumer

5.3.4 Communication Modes on Application-Layer Protocol Suites

The entire task when broadcasting native XML metadata is the building of a similar tree structure at the client and server sides. The server side means BSP, ISP, or SP or interservice-provider networks. In principle, textual or binary updating can be based on single updates, streaming, event or messaging mechanisms. Providing different communication modes enables more advanced communication modalities and intra- or interstream synchronization of metadata to multimedia content assets. The two protocols utilized in our architecture are based on SOAP and MPEG-7 BiM.

This section describes how metadata can be transmitted and synchronized, and states which applicable communication models are available. Different communication modes, where synchronization is especially pointed out, are also addressed. Other relevant characteristics of metadata communication models focus on *versioning*, *security and reliability*, *payload characteristics*, *modality* and *delivery scheme*. Table 5.1 gives a brief overview of those communication modalities.

Special focus is given to synchronization on different levels: where on the lowest level single continuous media stream synchronization takes place (see Table 5.2). Synchronization of multiple streams is done on the stream layer. The entities of operation are multiple or single streams that are synchronized with each other or with additional metadata. More significantly, the object layer emphasizes meeting deadlines as specified, to produce a presentation fulfilling several temporal constraints. The transaction layer is asynchronous and relies on user inputs for offering advanced feedback facilities. Still, preparation of multimedia material, such as applying video segmentation, requires longer time-spans, therefore when dealing with synchronization more soft-deadlines have to be met.

SOAP offers special handshake models [102]:

- fire-and-forget to single/multiple receivers,
- response/request schemes,
- remote procedure calls (RPC),
- requests with acknowledgments,
- request with encrypted payload,
- third party intermediary,
- conversational message exchange,
- message header and payload encryption,
- communication via multiple intermediaries,
- asynchronous messaging,
- encapsulation of non-XML data,
- multiple asynchronous responses,
- increment parsing/processing of SOAP messages,
- event notification, caching, routing, tracking, catching with expiration,
- quality of service,

Table 5.1. Metadata protocol types and their characteristics (see also Table 5.2)

	Feature	Description	SOAP	BiM
A	modality	single updates	♠	♠
		streaming	♠	♠
		textual	♠	♠
		binary	♠	♠
		event-based	♠	♠
		messaging	♠	no
B	synchronization	media layer	no	♠
		stream layer	no	♠
		object layer	no	♠
		transaction layer	♠	♠
		human layer	no	no
		material preparation layer	no	♠
C	versioning	protocol-based	♠	♠
		encapsulation-based	♠	♠
		incremental	no	♠
		continuous	no	♠
D	security and reliability	protocol defined	♠	♠
		acknowledgements	♠	♠
		encrypted payload	♠	no
E	payload characteristic	XML native payload	♠	♠
		RPCs	♠	♠
		other payload types	♠	no
F	delivery scheme	push	♠	♠
		pull	♠	♠
		push&pull	no	♠
G	routing facilities	intermediaries	♠	no
		routing capability	♠	no

- versioning of updates on protocol layer and
- attachments.

5.3.5 Simple Object Access Protocol (SOAP)

SOAP is an open standard to enable the exchange of lightweight XML messages within a distributed service pool. SOAP's principal use scenario is acting as an exchange protocol for multimedia assets in the form of XML metadata, serialized objects and attachments. Serialization means converting objects, their states, method calls and return values into textual format to obey persistence and unified object lifecycles. Additional object data, such as images, documents, etc. can be MIME encoded and attached to each message. SOAP message can be exchanged either directly or over intermediaries, whose purpose is mainly to forward messages to their ultimate destination. A yellow-page-like discovery service allows automated service lookup and discovery.

Table 5.2. Temporal synchronization models (in [24] extended by [125])

		Entity of Operation	Example	Characteristics
1	media layer	single continuous metadata stream	PES BiM application stream	device independent interface of operations and en-/decapsulation processes
2	stream layer	group of media streams, single media stream	MPEG-2 DVB stream, whole program incl. applications and content	interstream synchronization and intrastream synchronization
3	object layer	temporal synchron. specification as input	multimedia presentation, hyperlinked TV including object information	intra-/inter stream synchronization based on lower-layer calls
4	transaction layer	higher layer Internet protocols (HTTP, etc.)	XML-based eBusiness	asynchronous / synchronous
5	human layer	human interactions	activation of hyperlinked TV	asynchronous
6	material preparation layer	preoperational entities	extraction of metadata	soft deadlines

The functionality of XML content is either for the invocation of RPCs, or for pure communication purposes. Responses, requests and faults are encapsulated within textual XML descriptions. For more sophisticated exchange of binary content, the attachment facility enables carrying more complex messaging elements, such as compressed or uncompressed files, images, etc. Lightweight XML messages in a distributed service pool can therefore be easily exchanged over existing transmission protocols on the application layer. This enables easy protocol binding with HTTP, HTTPS, FTP, SMTP, etc. and therefore is network firewall-friendly and can be mostly accessed without restrictions in most environments.

Referring to simple XML over HTTP protocols, SOAP packets belong to the text-oriented exchange protocols, where serialized XML code is exchanged between nodes in a distributed service pool. Capabilities for attaching binary message parts (e.g. images) in MIME format are foreseen. The content of the message can be either for simple message exchange or the invocation of parameterized RPC calls that return their result to the requestor.

The Structure of a SOAP Message

Similar to packet-based network protocols, such as TCP/IP, SOAP's are in principal packets that are bound to other protocols, such as HTTP. SOAP differs in naming packets as envelopes, which wrap header and payload data. The *header* gives general information about the structure, encoding and authentication of the message. The header is immediately followed by the body of the message, which has payload functionality and contains the actual message data.

In more detail, four parts are present in each SOAP message: the header of the binding protocol, in whose payload the message is encapsulated, the *envelope* for wrapping messages, *data serialization (encoding) rules* for data type instances and a *remote procedure calls/response* representation. Similar to packet-switched networks, each SOAP envelope contains header information as the first immediate child. Figure 5.7 shows the general SOAP message structure and Table 5.3 one instantiated example message as illustrated in Fig. 5.8.

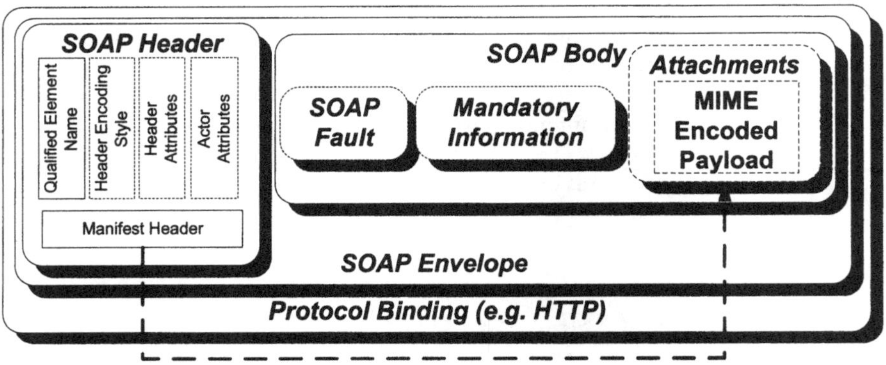

Fig. 5.7. SOAP message schema representation (adapted from [160])

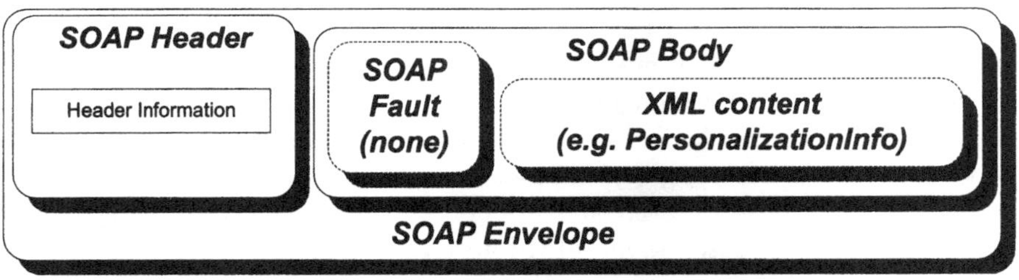

Fig. 5.8. Instantiated SOAP message example (adapted from [160])

Building a SOAP Message Header

To identify by which rules the envelope has been formed, the SOAP header contains a *fully qualified element name* in the form of a namespace URI. In practice it is a reference to an XML schema, describing the envelope structure. To maintain robustness and semantics of a message, the recipient has to know how to process the message. Three header entries (encoding style attribute, header attributes, actor attribute) are utilized for this purpose: the *encoding style attribute* holds information about those serialization rules that have been applied (e.g. `http://schemas.xmlsoap.org/soap/encoding`). The optional *header attributes* allow the encapsulation of how a recipient or intermediary of a SOAP message should process the message. This helps to maintain message semantics, error-free decoding and more robustness during messaging sessions.

For each header attribute an entry identifies its mandatory or optional usage. More relevant for intermediaries is the *actor attribute*, that clarifies how intermediaries process the message, and which header information is required at the destination of the message.

Structure of the Message Body

As previously mentioned, the SOAP body element contains the essence or mandatory information to be exchanged between two nodes. It is also identified by a fully qualified element name in the form of a namespace URI or local name, but can contain arbitrary data, whose structure is compliant with the encoding style stated in the message header.

Optional namespace qualification entries support the message recipient in validating the metadata payload structure of the message. In principle the message body can be utilized either as a carrier of data in the form of RPCs, XML, attachments or for error reporting. At the recipient side marshaling supports the process of converting SOAP message body entries to objects for further actions to be taken upon arrival. Table 5.3.5 shows several elements available for the creation of SOAP body elements.

Extended SOAP Capabilities

SOAP-based services provide extended capabilities in combination with other protocol types. Table 5.5 provides an overview of those features. The following section briefly points out their strength and use-scenarios in a deployment environment and gives references to relevant sources for further reading.

- *SOAP attachments* rely on existing standardized mechanisms for carrying base-64 binary data. Each document consisting of a root resource — such as a tree representation of XML instances — may contain distinct binary subelements. *Multipart/related MIME* [121] describes how to build documents consisting of arbitrary binary or textual elements and relations

Table 5.3. Instantiated SOAP message

```
<?xml version="1.0" encoding="UTF-8"?>
<SOAP-ENV:Envelope
   xmlns:SOAP-ENV=http://schemas.xmlsoap.org/soap/envelope/
   xmlns:xsd="http://www.w3.org/1999/XMLSchema"
   xmlns="http://schemas.xmlsoap.org/soap/envelope/">
   <Header/>
   <SOAP-ENV:Body>
      <isd:getRecomendation xmlns:isd="urn:RecommendationFetcher"
        SOAP-ENV:
           encodingStyle=
           "http://xml.apache.org/xml-soap/literalxml">
         <e type="interface org.w3c.dom">
            <PersonalizationInfo>
               <AvailableContentList>
                  <Programme>
                     <ChannelName>TV 1</ChannelName>
                     <Title>Morning News</Title>
                     <Genre>news</Genre>
                  </Programme>
                  <Programme>
                     <ChannelName>TV 2</ChannelName>
                     <Title>Formula 1</Title>
                     <Genre>sport</Genre>
                  </Programme>
               </AvailableContentList>
               <Consumer>
                  <Name>Samuli</Name>
               </Consumer>
            </PersonalizationInfo>
         </e>
      </isd:getRecomendation>
   </SOAP-ENV:Body>
</SOAP-ENV:Envelope>
```

between them within and outside the document's scope. In combination with *Universal Resource Identifiers (URIs)* [120], MIME referencing provides a powerful mechanism to identify each element contained in a document. SOAP attachments are based on this well-introduced mechanism. The SOAP message which is the root node of processing is referenced from the SOAP message-carrying protocol as the primary SOAP message. Each additional attachment, which can be either of binary or textual format, is uniquely identified by a content ID and can be referenced from the primary SOAP message. Binding SOAP messages that include attachments to other protocols (such as e.g. HTTP) requires setting content identification fields to MIME encoded type (e.g. `Multipart/Related`).

Table 5.4. Essential SOAP body elements (as based on [160]

Body element	Subdatatypes	Example(s)
simple types	int	`<personNo>10</personNo>`
	float	`<price>20E+1</price>`
	string	`<name>Hugo</name>`
	negativeInt	`<temperature>-3</temperature>`

	enumeration	`<element name="W3CStandards">`
		`<simpleType base="xsd:string">`
		`<enumeration value="XML"/>`
		`<enumeration value="SOAP"/>`
		`<enumeration value="HTML"/>`
		`</simpleType>`
		`</element>`

	array of bytes	`<arrayValues xsi:type="SOAP-ENC:base65">`
		`ddfFFeDf779CDe`
		`</arrayValues>`

polymorphic accessor		`<amount xsi:type="xsd:int">3</amount>`

compound type	compound values structs references to values	`<element:STB>`
		`<display>TV</display>`
		`<decoder>STB</decoder>`
		`</element:STB>`

	arrays	`<perfValues SOAP-ENC:arrayType="xsd:int[2]">`
		`<n>300</n>`
		`<n>100</n>`
		`</perfValues>`

	partial arrays	...
	sparse arrays	...

	generic compound types	`<abc:digiTV>`
		`<Monitor>Plasma Screen</Monitor>`
		`<SetTopBox>`
		`<Decoder>Satellite TV</Decoder>`
		`<Color>black</Color>`
		`</SetTopBox>`
		`</abc:digiTV>`

RPCs		`...`
		`xmlns:xsi="http://www.w3.org/XMLSchema-instance`
		`...`
		`<isd:getRecommendation`
		`xmlns:isd="urn:RecommendationFetcher" SOAP--`
		`ENV:encodingStyle="http://.../soap/"`
		`encoding/">`
		`<user xsi:type="xsd:string">Hugo</e>`
		`</isd:getRecommendation>`
		`...`

- *WSDL:* Where SOAP belongs to the network protocols on the application layer, WSDL describes how to exchange messages and which capabilities clients or servers must have to understand those messages. WSDL focuses on the deployment description of SOAP-based messages to obey higher interoperability between nodes. This requires the provision of rules and guidelines for how to communicate between nodes that have to be followed in order to achieve a certain goal during a SOAP-based transaction session.

Table 5.5. Extended capabilities of SOAP as transmission protocol

Capability	Description
SOAP attachments	SOAP enables the attaching of MIME encoded data structures
MPEG-7 metadata encapsulation	encapsulation of MPEG-7 DDLs, DSs and Ds
MPEG-21 metadata encapsulation	encapsulation of MPEG-21 digital items (DIs)
generic XML encapsulation	carriage of namespace qualified XML schemas and instances
protocol binding	easy binding to lower-layer protocols, such as HTTP, IP etc.
WSDL	the *Web Service Deployment Language* describes message exchange procedures
yellow pages service	provision of lookup mechanisms for finding and linking services to each other

- *Generic XML encapsulation:* SOAP messages enable two modes for encapsulating other metadata types: payload-based as attachment and payload-based RPC of whole documents. The simplest facility to transmit generic XML content is attaching it to the message MIME encoded, by setting the MIME header to `type=text/xml` and the MIME message body to `content-type: text/xml`. Adding XML documents as parameters to RPC calls requires an additional entry in the SOAP message header to notify the client of how it should process the message. A simple namespace entry in the form of an URI, qualifying the RPC namespace content is sufficient (e.g. `xmlns:e="target-URI" SOAP-ENV:mustUnderstand="1"`). RPC calls also require a qualified namespace of the parameters to transmit. It is the same `target-URI` as utilized in the header format. As in an RPC call, only a restricted set of SOAP-encoded types can be carried. It is only capable of delivering simple generic XML content types. Therefore sending MIME attached XML content types is preferred.
- *MPEG-7 and MPEG-21 metadata encapsulation:* Generic XML encapsulation techniques can also be applied for transmitting MPEG-7 and MPEG-21 based metadata. It differs only by replacing the `target-URI` with the MPEG-7 (e.g. `urn:mpeg:mpeg7:schema:2001`) URN or MPEG-21 (e.g. `http://www.w3.org/2001/XMLSchema`) URI. This is a sufficient solution for transmitting this type of metadata. Sending MPEG-7 and MPEG-21 as attachments requires no changes of the SOAP message as done by generic XML encapsulation.
- *Yellow pages service:* Finding other SOAP-based services and their definition for interchanging content formats as defined by WSDL is part of

yellow page service providers. A lookup search mechanism provides access to a database of entries, relevant for a special service.

5.3.6 Streaming Binary XML — The MPEG-7's BiM

The objective of streaming binary XML is to provide a sophisticated solution to obtain metadata in binary format. A current approach is defined within the MPEG group, which specifies MPEG-7 Systems. Within the scope of this standard, terminal architecture, multiplexing and synchronization of binary MPEG-7 structures is defined. Binary streaming of XML is not essentially related to the MPEG-7 descriptor language. It is a generic protocol for any type of XML related metadata. The goal of MPEG-7 BiM's representation is to transmit complete metadata trees, or some subtrees to the terminal architecture. Mechanisms for tree reconstruction, transport packet format, synchronization, delivery, BiM and mapping textual metadata representations to binary form are defined. The key features of this transmission protocol are as follows:

- *incremental update* of parts of the metadata tree: complete transmission of metadata trees is obsolete, due to the incremental update of only parts of metadata trees. Therefore complete data structures do not have to be transmitted each time;
- *low bandwidth requirements* due to high compression rates of textual metadata representations: compression rates of pure textual information are rather high;
- *smaller parsing times* due to the binary format and partial metadata trees: parsing of complete data structures is obsolete, as only subtrees are transmitted. They can be parsed at less computational costs in terms of memory and performance;
- *facilities for metadata validation*: schema transmission via BiM enables validation of XML streams at the decoder side;
- *enables enhanced decoder/encoding features*: lookup table mechanisms at the client side enable higher compression rates, as elements are referred to via lookup entries. Synchronization has to be done mostly by utilizing mechanisms of lower layer protocols, but their invocation is not needed directly;
- *supports pull and push/pull schemes*: both schemes of transmitting data enable the utilization of this protocol either for feedback or broadcast channel use;
- BiMs are a *sophisticated solution for storing metadata*: high compression rates help to save storage space;
- *encryption*: the binary stream can be encrypted with a 128 digital signature key.

MPEG-7's BiM covers the presentation (compression) layer and is built on other protocol types of the ISO/OSI multilayered protocol stack. The unit of transmission is an *Access Unit (AU)* consisting of the header and payload data. AU streams are encapsulated in other lower-layer protocols, which may vary from application to application (e.g. IP, MPEG-2 TS, RTP). This fact and the nature of binary metadata representation enable upstream and downstream capability. Therefore its utilization in push or pull/push environments is an intelligent usage scenario. Depending on the application's needs, metadata is losslessly compressed into its binary formats. In some cases it might be relevant to leave out comments or other nonessential parts of a metadata tree, therefore BiM streams might be not a bijective representation of textual metadata descriptions. The BiM encoder traverses the metadata tree and compresses the relevant textual descriptions into its binary format: elementary streams, either for metadata schemas or instantiated metadata. Schema streams are utilized at the decoder side to verify instantiated streams on their correctness.

At the decoder side, elementary streams are decapsulated, decompressed and transcoded into a format relevant for the application. Elementary streams containing schema streams are decompressed and utilized for further definition and validation of metadata instance streams. Each AU is converted back to its textual representation, but mostly into a tree-like data structure. Unfortunately in most cases the decoder has to rely on synchronization and timing mechanisms covered by lower-layer protocols and is not included rigidly into the AU mechanism. Basically the decoder acts as a catalyst for decompressing, validating and transcoding binary representations into a structure required at the application layer. It might be relevant to obtain textual descriptions, invoke lookup tables for resolving element names or directly construct a metadata tree from its binary format.

Why not simply zip a metadata file and send it to the client? A binary transmission, as defined by MPEG-7 Systems, differs from simply compressing an instantiated metadata file. First, continuous updating of metadata subtrees is possible. Second, transmission can be handled more flexibly due to the fact that the binary format is streamable and navigable. Furthermore, for processing tree elements the overall metadata instance does not have to be parsed, only the relevant sections as required by the application.

Access Units (AU), the "Network Packet" of MPEG-7 BiM

The key feature is that each AU represents a useful chunk of a metadata tree, holding a complete set of data to reconstruct a complete metadata tree, or some parts of it. Each AU maps textual metadata descriptions one-to-one to its binary representation in its payload and adds navigation information into its header. Each header contains entries that uniquely identify at which position at the client side a metadata sub-tree has to be added. Each AU maps textual metadata descriptions bijectively to its binary representation. The AU stream is continuously parsed at the client side to update the internal representation of metadata trees.

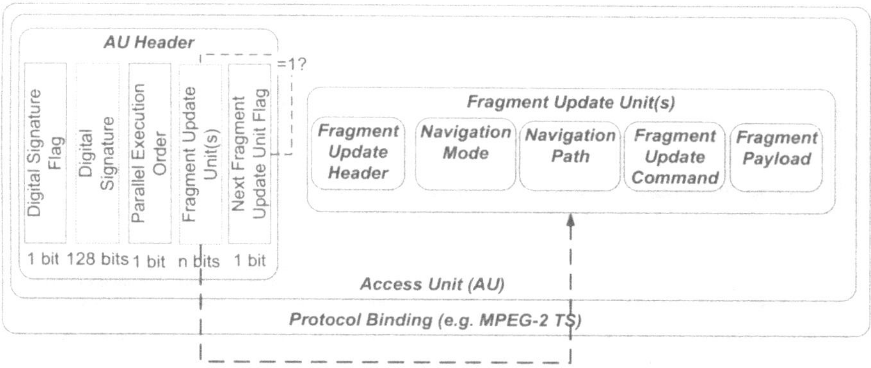

Fig. 5.9. BiM data packet consisting of *access units (AUs)*, *fragment update units (FUUs)* as payload and the carrying protocol (e.g. MPEG-2 TS) based on MPEG-7 standards

To decode the binary metadata stream several entities of MPEG-7 systems are relevant: First, the *decoder configuration*, containing information about the schema URNs, navigation mode of changed child elements contained in the stream, digital signature for the decoder configuration and system relevant information. Second, the AU itself contains not much header information, such as a digital signature for encryption, an execution order and the more complex structured payload as a *Fragment Update Unit (FUU)* which is the third relevant entity. The FUU carries the actual binary representation of metadata structures.

Table 5.6. The essential parts of a *fragment update unit (FUU) as based on MPEG-7 standards*

Structural Element	Purpose	Content
fragment update header	MPEG-7 definition identification and packet length alignment	
navigation mode	global location where nodes have to be added	absolute to root or current top-level element, or relative to current node
navigation path	local location where nodes have to be added	tree branch codes (TBCs) schema branch codes (SBCs)
fragment update command	modality of adding subtrees	add, replace or delete
fragment payload	encoded metadata	textual or binary representation

Each FUU starts with a top-level element, which can be either the root element of the whole document or one subtree element of the document tree. The navigation mode defines the global location of the subtree encapsulated in the payload in the overall metadata tree. For adding, removing or replacing elements the navigation path refers locally to the location as determined by the navigation mode. Which action has to be performed is determined by the navigation mode. *Tree branch codes (TBCs)* help to determine the absolute position dependent on the navigation mode. Thus, an absolute path relative to the root node is referenced by its TBCs (e.g. the first encoded element in the payload is the first child node of the document root element). TBCs are tables of positioning codes for child elements dependent on the position of its parents.

Processing BiM Messages

In the following the steps to decode MPEG-7 BiM streams are stated briefly. The process of decoding MPEG-7 BiM streams starts by setting up the decoder by utilizing the decoder configuration for initializing decoder instances, decryption facilities and system resources. The input of the decoder can be either textual or binary representations of schema or metadata instance streams.

Schema streams are simply utilized to validate metadata instances by their canonical equivalence. Upon validation the application gains access to the required data for further processing. What it does with the data depends on the service type. The decoder implementation itself is basically a finite-state automaton for parsing and verifying metadata trees. Figure 5.10 shows the principle of AUs and their binarization processes.

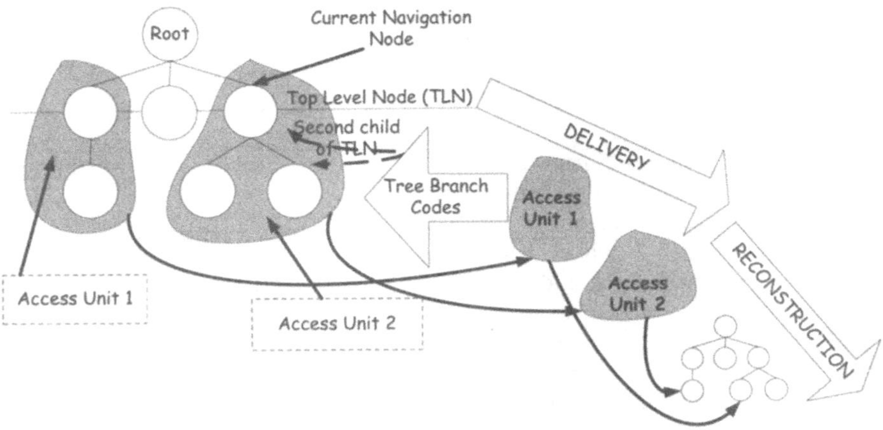

Fig. 5.10. Relations between metadata trees and their binary representation

5.4 Consumer Devices

Nowadays market-launched multimedia home equipment, different communication models between devices and several user interfaces are already available. Their complexity is obvious. But multimedia assets have to be retrieved, modified, displayed and used interactively. This implies certain unification and development of standard architectures for performing equipment integration. This section focuses on digital TV equipment as an access point to integrated multimedia home services (e.g. light control, heating system control and automated food ordering if the fridge is empty).

Different predictions of what the service access point will be at home exist. The question is what the multimedia hub at home acting as a front-end to services will be like. Hi-Fi equipment manufacturers state that the Hi-Fi at home will be it; digital TV manufacturers offer an STB acting as access point; game console creators think they will cover this market; and some still believe it will be the PC or mobile equipment such as PDAs or mobile phones. What will be the major market-deployed equipment decided by the consumer in coming years? Within this book we rely on digital TV equipment as an unique access point to multimedia home services [169]. A very abstract model of digital TV-based consumer equipment is presented in Fig. 5.11. It shows the principal building blocks of a multimedia home equipment digital TV service hub.

The key questions in the development of services for consumer devices are: Which user-interfaces are adequate for the end-user? Which additional equipment is essential to experience TV in an enhanced way? Which presentational content possibilities exist? Which equipment is essential? How will user-interfaces be designed and laid out to guarantee that all age groups have appropriate access to content? New input devices and methodologies are required as television is a collaborative experience by nature. This requires new input and presentational methodologies over devices that enable the end-user to interact with content without disturbing others. Figure 5.12 shows the client rendering pipeline for MHP-compliant consumer devices.

PDAs and Bluetooth are existing solutions for advanced interaction devices. But pointer devices and enhanced remote controls could also lead to major enhancements in this environment. Not only the interaction between the end-user and the TV device can be considered converging digital TV to a home-network solution is also a basic issue. This enables accessing home services, such as the heating system, wireless and wired home-networked devices and other networked facilities in an enhanced way.

The involvement of different input devices and new methodologies is possible. Pointer devices, by which the end-user can interact with content by simply pointing a laser-pointer device at the TV screen, promise new emerging interaction facilities. During a shared access of TV content, PDAs assist in individualizing the television experience. They might act as certain add-on devices, with which content add-ons can be viewed or interacted.

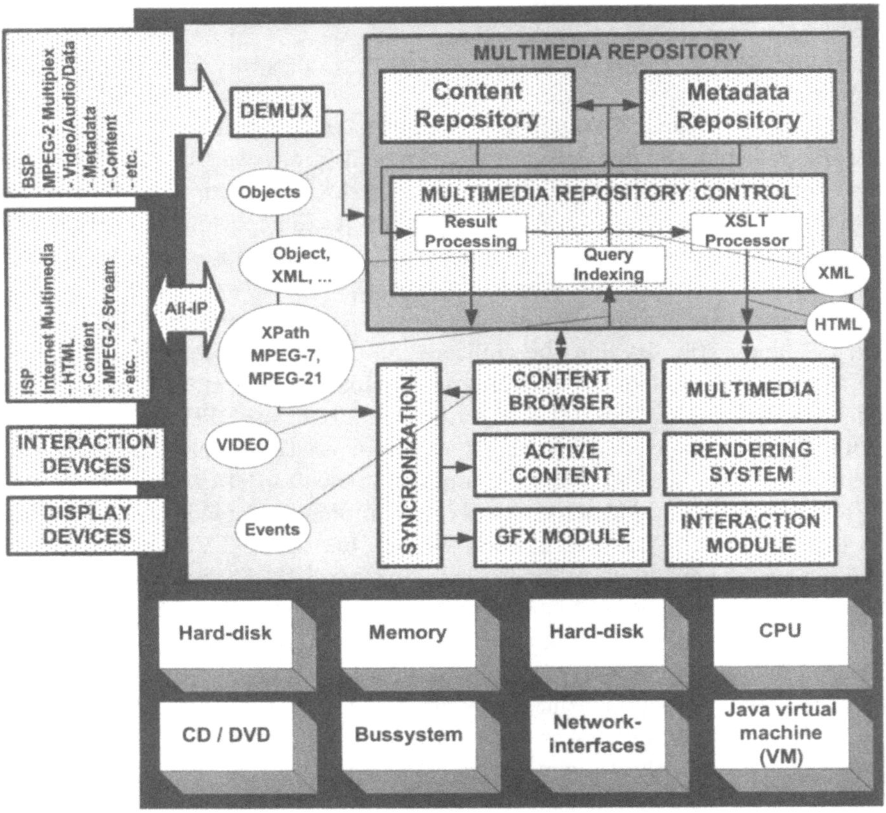

Fig. 5.11. General client architecture

Presentational enhancements, such as new output devices or multiple screens also allow very promising applications. The communication protocol might be Bluetooth or infrared between the box and the device. But also the gaming industry delivers more advanced and high-performance devices in the forms of game consoles, head-mounted displays, or TV joysticks.

Each device has to be integrated in a certain manner, over adequate communication protocols. Smaller gadgets, such as Bluetooth interconnected e-Pets and game-boards. give birth to new services and eCommerce solutions. The major concern is gaining access to the local TV from anywhere, at any-time and in any form. These facilities provide a more advanced television system, synchronization, transmission and user profile management for accessing equipment remotely. To realize software-based services, MHP has standardized the principal *application programming interface (API)*, as shown in Fig. 5.13.

Content presentation, enhancements of user interfaces, its graphical presentations and advanced sound output schemes let current digital TV converge

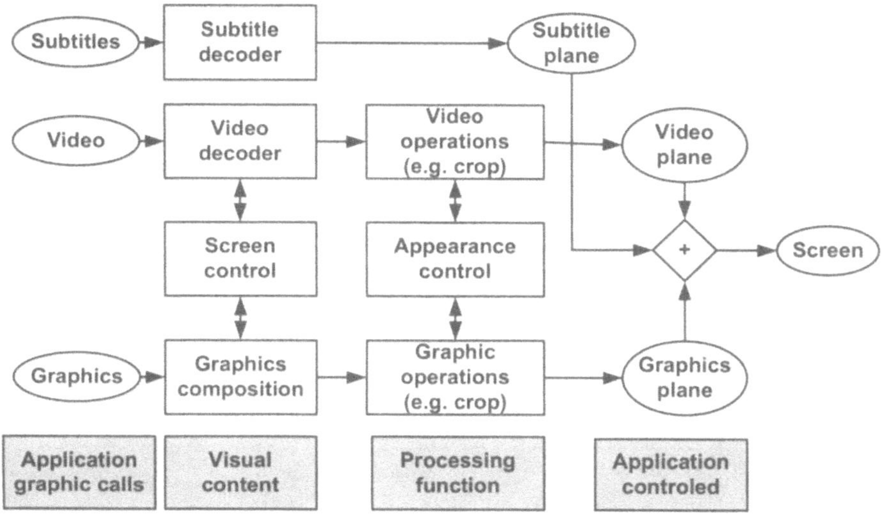

Fig. 5.12. Client rendering pipeline (adapted from [68])

to a complete interactive show. It is difficult to predict any future developments and new emerging devices. This chapter just presented some snapshots of current systems and solutions. The future and consumer behavior will show which ones will be the real breakthrough in digital TV.

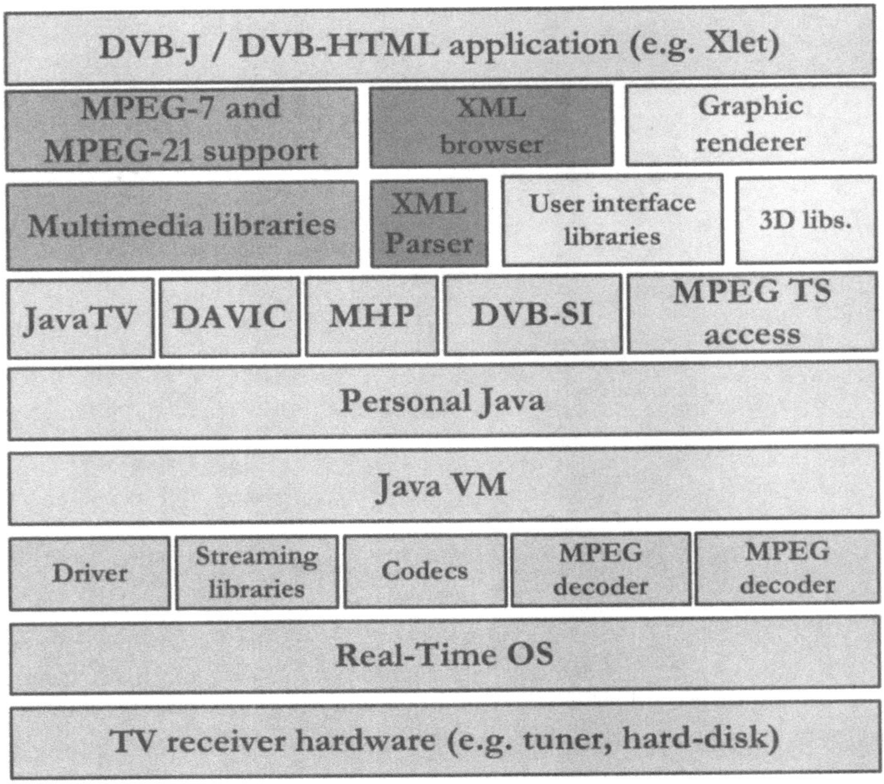

Fig. 5.13. Client software architecture

Part II

Application

6

Innovations in Digital TV

Digital TV is more than just compressed video, improved picture quality and human user interface design. Consumers are embedded in and surrounded by services with transparent software and invisible, easy-to-use devices, wherever they go, whenever they want. Simply put, digital TV is "*new — wireless — immersive — ambient*".

The consumer is merged into a digital service space that tells its narratives either passively or actively involving consumer interactions. This is the basic concept to be understood in developing novel services on this emerging platform. It is important to note that the term "service" refers to two different things. DVB standards describe services as video streams, compiled programs, etc. For use-scenarios services refer to complete end-to-end solutions such as applications.

In this chapter highly visionary and innovative scenarios, potentially utilizing the DBI methodology, are introduced. The subchapters are divided into general topics, briefly introducing possibilities within this application area and into specific topics representing our ideas of innovative services in digital TV:

- introduction of digital TV as a disruptive innovative technology,
- illustration of the digital TV paradigm shift,
- establishing a consumer-oriented service model,
- definition of digital TV as a narrative,
- categorization of services with the help of the narrative cube,
- enumeration of novel metadata-driven services and
- pinpointing related technological challenges.

6.1 Paradigms in Digital TV

Digital TV will create a paradigm shift in broadcasting. Rigid, inflexible and mostly publicly driven companies will feel this shift very strongly. Broadcasting will change and a fundamental shift towards interactivity and Internet

Table 6.1. Digital TV services

Service Type	Description
PC Migrated Services	
communication	E-mail and newsgroups, instant messaging, chat, teleconferencing
collaboration	gaming, content synchronized chats
Standard Digital TV Services	
informational	TV portal, EPGs
regionalized	regional weather, regional news, merchandising
personalized	custom news, advertisements, personalized EPGs, automatic video recording
interactive	game shows, interactive advertisements, eKnowledge platform, knowledge visualization, transactional services, access point to the digital smart home, user authentication (smart cards, conditional access, fingerprints, voice identification), payment schemes
Visionary Services	
perceptional	active content and content manipulation, perceptive digital items, adaptive content, personalized characters and actors
collaborative	digital communities, contextual computer games, TV as an artificial companion
narrative	interactive narratives, knowledge socialization
intelligent	smart TV, observation of user habits, user group identification

convergence will be clearly coming. Looking back to the most recent years, the most meaningful activity was the development of the Internet culture. Consumers are enabled to interact with content. TV remained a passive medium without any — or very limited — interaction possibilities. This is a place for newcomers in the business to create their own markets. Broadcasting, or in a bigger sense broadcast multimedia, faces these facts: convergence, increased digital perception and awareness, creating meaning in digital content and digital communities.

Convergence evolves in three lines: *device convergence*, *distribution channel convergence* and *service convergence*. Convergence means the creation of the digital multimedia home server. The digital home server is the device that plays music, sings songs, controls the fridge, stores digital holiday movies, speaks to its user and so on. Many industrial partners compete for this device. It is currently not known whether it will be a PC, game console, Hi-Fi equipment or a digital TV STB. Who will be the winner is still an open question. Convergence also means convergence of distribution channels: wireless, wired, Internet, broadband or digital broadcasting channels. Different channels will substitute for each other and provide an alternative to distribute multimedia assets to the consumer. Therefore digital TV might not only be distributed via DVB-T, DVB-S or DVB-C; its biggest competitor is the Internet with its increasing bandwidth. DSL technology already enables the consumer to enjoy a low-quality digital TV show.

Service convergence is also a big issue. Service convergence implies the fact that multiple platforms which provide similar services for the consumer are interconnected, accessible from anywhere at any time and have different features and characteristics in providing this service. A very simple example is a video recording device connected to the Internet. It could be programmed from a PC, a mobile phone, an office PC or a remote control. Service convergence addresses the issue of delivering bouquets of services from a broadcaster to different consumer platforms over different distribution channels.

Once the service is distributed, it can be accessed from different devices and contribute to a *service space* (service grid) for the user. Therefore the service provider builds a service space for the consumer over various distribution channels. The consumer will also select the most convenient space. Current digital TV equipment lacks mostly in performance. Users might switch to PCs in order to obtain the latest TV program and use the PC as an interface for programming the video recorder device. Service convergence therefore means the creation of a service grid of interoperable applications, where services are accessible anytime, anywhere, anyhow and available to the user.

Our paradigms for digital TV are:

1. digital TV applications are disruptive and digital TV is at the beginning of its S-Curve;
2. the producer creates more than simple movies, he creates a service space or fictive TV universe for the consumer;
3. consumers, content creators, narrative, interaction and multimedia assets belong to the service space;
4. the narrative cube is a model for creating semantically and aesthetically improved broadband multimedia story-telling environments;
5. considering the correlations among interactivity, the narrative and multimedia assets within the fictional TV universe delivered to the consumer;
6. digital TV must not only be seen as broadcasting; it is a convergence between distribution channels (e.g. Internet, DVB-S, DVB-T), hardware platforms (e.g. hi-fi equipment, STBs, PCs), multimedia home equipment (e.g. digital camera, etc.) and the content that is presented (e.g. computer games converge with digital broadcasts);
7. natural perception of content and awareness of computer-mediated communities.

6.2 Digital TV Es Innovative and Disruptive

Digital TV is a new technology and the first business revenue will surely come from well-known commercial applications such as eShops, TV fees, subscription TV, video-on-demand, chatting, etc. There is more to digital TV than the deployment of standard applications. Three forces can be identified: estimated *consumer needs* available through consumer research, *disruptive technology* arising from universities and R&D labs and finally consideration of what digital TV really means and which *new paradigms* are to be considered in the world of digital broadband multimedia.

Commonly known business models such as pay-per-byte, pay-per-view, pay-per-service, etc. will surely not be outdated. On the other hand, consumers like flat-rates and TV is commonly a free environment, where only some monthly payments have to be made. Business models in TV will evolve around the following issues:

- development of applications, content and system infrastructure delivered to BSPs, SPs, SEs, etc.;
- flat-rate for consuming content in addition to some pay-per-service models;
- increased competition for higher audience ratings among BSPs due to more differentiated service offers;
- advertising as the top-revenue model even in digital TV.

Dealing with disruptive technology is a complex and difficult task. Many questions arise and many different opinions exist. Three basic issues are considered within the scope of this section: *phases of disruptive technology*, the *S-Curve*, *disruptive marketing strategies* and the *10X Factor*.

6.2.1 Phases of Disruptive Technology

Disruptive technology undergoes three principal phases of evolution (paraphrasing [136]):

1. a *proof-of-concept* phase gives technology insights for design, concepts and product development. Nobody will claim that this idea ever works every time and general suspicions are all present. The common opinion is that the "researcher in his ivory tower invents something without evident value". Opinions about the "uselessness of the idea" are excellent indicators of new emerging disruptive technologies, as its concepts are not yet understood;
2. the *limited application* phase follows the first phase. Common opinion changes towards considering "why utilize the new and unproved concept when the old paradigms work perfectly". The idea is understood and finds potential market segments since it will be recognized as useful, applicable and a unique solution. Still, a final breakthrough has not been achieved;
3. it is a large step from limited application towards *widespread applications* on the market. The market segments have been found and the common opinion changes to seeing it as a competitor to well-established technological solutions. Industry accepts the new technology as a proven, reliable and modern solution to an old problem.

Example 6.1 (TV-Anytime is disruptive). The use of digital TV receivers as personal data recorders was first considered by TV-Anytime as a potential technological novelty. Together with the TV-Anytime metadata model, a novel technology for consumer devices was born. TV-Anytime developed through the three phases of disruptive technology: starting from an R&D project, limited application in some demonstration implementations and now widespread applications becoming part of the DVB standardization efforts.

6.2.2 S-Curve and Digital TV

The S-Curve model complementary to the phases of disruptive technology [11] is presented in Fig. 6.1. Where phases of disruptive technology focus more on one concrete idea, the S-Curve model describes overall industrial development, phases of novel technology branches, maturity, innovation lifecycle patterns and product development over time. It supports whole organizations or branches of it to help them in relation to their current standing and the strategic decisions they have to make for future survival.

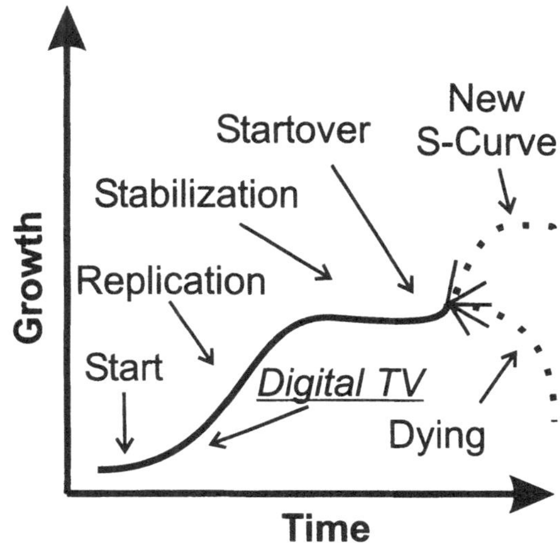

Fig. 6.1. The S-Curve after J. Abraham [11] adapted for digital TV

Any technological innovation starts with an idea or a product which is realized. It enters the replication phase, which means improving services and investments for making a business out of the idea. The idea has to be kept in the stabilization phase as long as possible, as revenues are made at this point in time. This is the point where businesses look for novel ideas and initialize a new S-Curve. At this stage products die or find a new start-over.

Strategically leaders need to know where they are in order to initialize novel innovation cycles to maintain the growth of their business [11].

Example 6.2 (digital, interactive TV in the S-Curve). Digital TV as such is an excellent example of being at the starting point of the S-Curve. It is a new innovative branch with new large revenue potentials. No market saturation or stabilization is currently in view. Many business leaders develop potential new ideas which are continuously improved. Examples are all-present: TV-Anytime, production systems, metadata systems, digital rights management systems, standards and file formats.

6.2.3 10X and Digital TV

Michael Porter [146] identified various factors in defining how competitive a company is. These factors are described and paraphrased in other works (e.g. [96]) on which this book is based. In principle, the fear of each business is the 10X force. The 10X force indicates the tremendous change in how business can be conducted [96]. The convergence between digital TV and the Internet is such an example and changes the way consumers are using their personal investments.

Figure 6.2 shows digital TV as a 10X change, by converging Internet and traditional broadcasting. Changes might include Internet broadcasting, third and fourth generation of wireless networks (enabling bandwidth of over 100 Mbps), consumer expectation of interacting and many other unpredictable ones. This clearly requires rethinking the impact digital broadcasting has on existing traditional revenue models. More potential competitors used to conducting business on the Internet enter the market.

Traditional broadcasters, which used to be publicly funded monopolies, face the existence of other distribution channels such as the Internet. Old revenue models based on monthly fees or advertisements might not be sufficient to ensure consumer loyalty. Additional interaction services, localized advertisements, fancy services, etc. are add-on to the actual content, but will be a major factor in keeping the consumers. Digitalization also means a chance for newcomers in the very exclusive club of traditional broadcasters. Newcomers have Internet experience and are less restrained by the traditional limitations of broadcasting monopolies. Streaming over the Internet does no longer require applying for expensive licenses or broadcasting frequencies.

Due to the novelty of digital TV it is rather difficult to estimate where it will go and which 10X forces exist. Figure 6.2 shows some potential forces and scenarios. The future will show where and how they will develop. But at least one thing is already clear: there is enough potential for disruptive technology.

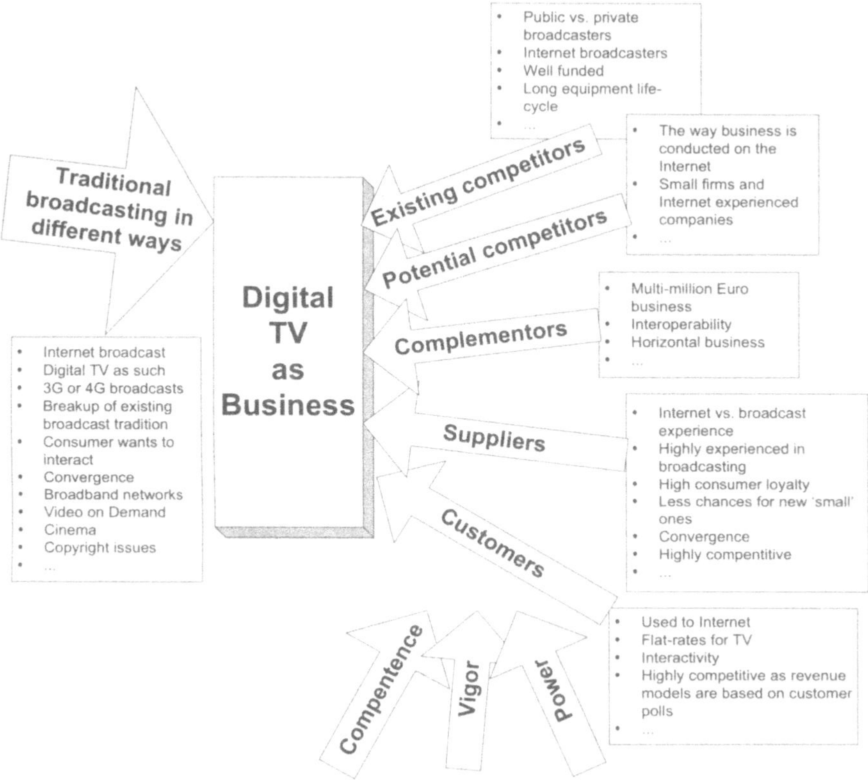

Fig. 6.2. Digital TV as 10X force

6.3 Asking the Consumer

Our statement is clear: asking the consumer which new applications he would like is the wrong way to approach new technology. He will answer that he likes to read e-mail, watch movies, listen to music, play DVDs, etc.

Consumer research will only work out for well-established new technological paradigms. No one would have predicted the success of typing messages on a clumsy mobile phone keyboard and sending them to one's friends. Nowadays SMSs or text messages on mobile phones have a large market volume and big revenue margins, since they cost service providers very little in production. The conclusion is that the consumer does not know which disruptive technology he wants. On the other hand, to leave consumer opinions out of the picture is also wrong. Consumer-centered service design is a major concern! But the consumer is the wrong person to answer the question about which *innovative* technology he would like to adopt. Figure 6.3 shows results from a consumer research study performed in Finland in 2001. According to this the most wanted services in digital TV were the news and weather, TV guides (e.g. EPG), regional announcements, movies on demand, information retrieval and e-mail.[1]

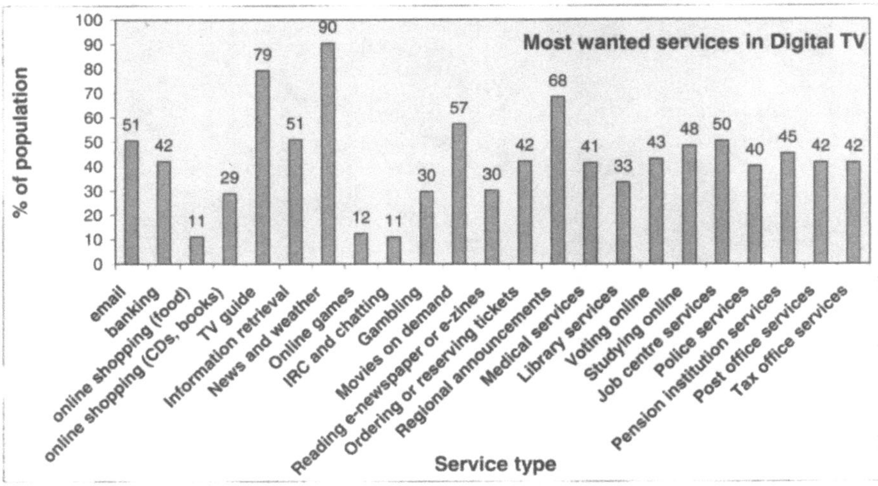

Fig. 6.3. Most wanted services in digital TV [162]

The *consumer-oriented service model* or *synthetic human–computer–human service model (H-c-H service model)* (see Fig. 6.4) is a trial for categorizing services in digital TV from the system user's viewpoint. Making technology invisible and moving the user into the digital world without changing perception from the real-world are its major concerns. The real-world acts as a reference point for how consumers interact, perceive, move the multimedia assets and how they come together in computer-mediated communities. Similar modalities are created in the digital world too. Especially consumer-oriented services in a digital space are not only technological challenges. The consumer finds herself in a service space. She perceives this place as it would be in the real-world. Therefore natural interaction, perception and awareness of things, communities and the person herself are not to be neglected.

Virtual reality (VR) already made a great effort in researching these issues from the perspective of surrounding people with a virtual, computer-generated environment. Human–computer interaction deals with the issue from the viewpoint of designing easy-to-use systems for consumers. Between the research fields there is a huge gap: the first one embeds the user completely into an artificially created world; the latter focuses only on devices, user-interface

[1] We would like to thank Sanna Leppänen from the TATU research group at the Digital Media Institute, Tampere University of Technology, Tampere, Finland in providing this material. The results of her research have been published in [162].

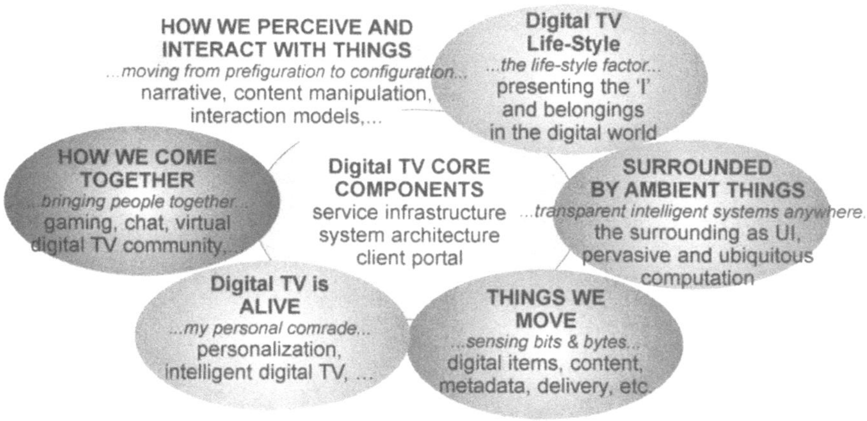

Fig. 6.4. Consumer oriented service model

design and empirical user studies. The question is, what lies between both research fields? The user is surrounded by devices, interaction models, narrative models, converging home leisure equipment and intelligent ambient devices.

Let us fill the gap with the description of the synthetic H-c-H model. The synthetic H-c-H model lets the "c", the computer system, be transparent or invisible to the consumer. Human–human interaction is the major concern, involving disappearing hardware and software. The synthetic H-c-H model is a human-centered categorization of services. This includes perception and interaction with digital things, establishment of digital communities and intelligent human-centered system design, surrounded by ambient technology and moving digital content in an intelligent way. The awareness of one's self and one's digital representation and presentation in the digital world including sociological factors (e.g. life-style, creation of digital communities, communication behavior models) are predominant. Within the scope of this chapter we describe this model in further detail.

Example 6.3 (chatting in digital TV). Chatting on the Internet is very popular nowadays. It is a matter of time until chatting services will be provided on digital TV platforms. Earlier chatting was realized via creation of a very limited presentation and awareness of its own in the digital world. But digital communities form and make use of the technology that is available. They use digital smilies (e.g. ";)") to show facial expressions and mood and nicknames for simple representations of their identity. The fact that people come together in the digital world and analyze chats from social points of view allows drawing relations between real-world and digital behavior. More advanced chat types involve 3D graphics for human presentation in digital worlds, as shown in Fig. 6.5.[2]

Fig. 6.5. "Habbo Hotel" creates three-dimensional chatting spaces (© Sulake Ltd.)

The question is now, what does this model have to do with metadata in digital TV? The answer is simple. Metadata supports the design of such systems due to its descriptive character:

- platform independent interoperability,
- adaptability of content to multiple end-devices,
- scalability of services (e.g. for handicapped persons),
- unified and generally applicable service-oriented data models,
- data analysis support for obtaining consumer data,
- presentational and representational data models and

[2] We would like to thank Sulake Corporation Oy, Helsinki as creators of the 3D chat "Habbo Hotel" for their screenshots. Habbo Hotel is an online gaming environment for teenagers. It's a place where teens can meet up to play games and develop their self-expression. Habbo Hotel provides a safe, rich and positive gaming environment, but it's the teens themselves who write the script. As "Habbos" they create their own character, room and virtual world, by interacting and playing with others (see `http://www.habbohotel.com` and `http://www.sulake.com`).

- rapid and fast document-oriented content development.

These are only some of the reasons for applying metadata for a more consumer-oriented design methodology. In this chapter, we show how to apply metadata models with the consumer-oriented service model. The core components of this model are existing technologies that have been described in detail in this book. Services are built on top of the technology and implement the consumer-oriented service model. The issue has been addressed by numerous research works [126].

6.3.1 How We Perceive and Interact with "Things"

Perception is the "the process by which an organism detects and interprets information from the external world by means of the sensory receptors" [53]. Multimedia assets manifest in many arbitrary forms, such as music, video files, e-mails and slide sets. Digitalization in television means the creation of a service space for electronic items, for their exchange and storage. The consumer perceives them in a more physical, thus "real" way. Simple examples are the myriad dead bits and bytes on hard-disks, where the consumer does not find anything anymore. He is a mass consumer of digital content and loses any emotional and personal binding to them. Adding perception and awareness creates personal binding and eases interaction with things in a natural way.

Perception evolves around interactivity, multimedia assets, structure and presentation. Within the scope of services the term "content" is used to refer to the multimedia asset. Presentation addresses the aspect of how multimedia assets are perceived by the consumer's senses. Adequate methods for rendering audio-visual components and stimulating senses are predominant. Furthermore, interaction means perceiving interaction devices (either artificially created ones in virtual environments or real-ones manifesting as physical entities) as objects for interacting with multimedia assets. Any methods altering the original multimedia assets towards a more perceptive form are allowed. As digital TV is a highly narrative medium, the narrative and its structure are key-factors for perceiving content. Narrative structures rather than static content types are perceived by consumers. This means the creation of a story flow and structure. Adequate interaction modes relating content, narrative flow and user inputs enhance the perceptual components.

From the technological viewpoint "perception of things" manifests in many different ways and relates directly to the multimedia asset types. Metadata is a key component for describing content (e.g. as MPEG-7 describes multimedia assets in a database). Metadata is a catalyst for representing and presenting multimedia assets. Therefore segmentation mechanisms for multimedia assets, metadata models containing perceptive components, audio profiling, metadata-based multimedia presentation languages (e.g. SMIL, SVG) and dynamic content descriptions are dominant for increasing perception. Figure 6.6 presents an example of visualizing audio and spoken content for people with a restricted sense of hearing.

Fig. 6.6. Sketch for visualization of audio content for people with special needs

Example 6.4 (increasing perception by using metadata for advanced visualization). An example of enhancing perceptive components is the delivery of metadata to consumer devices enhancing the visual representation of video content (e.g. exchange of actor's face or regionalization of advertisements).

Example 6.5 (active content elements for increasing the consumer's perception). Active content sees content as a dynamic, self-evolving and user-interaction-driven medium. Content or multimedia assets are not simple static, monolithic structures presented to the consumer. Real-time content manipulation or models to create new evolving content types especially belong to this category.

Example 6.6 (adaptive content). Services are distributed to the consumer anytime, anyhow and by her means. Adaptation of content enables the consumption, therefore perception, of media independent of its location. It is available throughout heterogeneous networks and scaled to the needs of the consumer. Metadata is the key-element to universal multimedia access as defined by MPEG-7 or in MPEG-21. It also enables handicapped persons to perceive content by the senses available to them as shown in Fig. 6.6.

6.3.2 How We Come Together

In general, there is human–computer interaction and human–human interaction and often human–human interaction is mediated by computers. Digital TV is available throughout all social levels and is a well-accepted device in homes. The question is, how digital TV can be used as a device for building digital communities, where the computer in human–computer–human interaction disappears or is as "transparent" as possible. This relates to group communication in cyberspace. The features are described in [152, 32]:

- *uninhibited behavior* including swearing and direct offenses;
- *depersonalization* implicating de-individualization;
- *reduced emotional behavior* in comparison to face-to-face meetings;
- *reduced self-reflectance and reduced fears of rejection* of one's own behavior and reduced fear of rejection compared to face-to-face meetings.

Digital TV and its application bring digital communities into every consumer's home. New technology decreases anonymity and self-reflectance. Techniques similar to face-to-face meetings enable communication and enhance the feeling of real presence in cyberspace. Being present in cyberspace and actively realizing this fact prevents depersonalization and uninhibited behavior. Due to the tremendous technological advancements such as high-bit-rate video streams and broadband networks, one's own presence in cyberspace is likely to increase. Since the computer as a medium or interface between humans is most likely to get more transparent, the perception of talking to a computer system decreases. During chats a simple nickname will not be sufficient anymore. Graphical representations or real video will be required. This is also true in creating communities of similar interests. They form spontaneously and have a common goal. Digital and real borders merge into each other, therefore perception and awareness of communities in digital form will increase. This also includes the search for communities with similar interests.

The technology behind digital community building services is rather complex and involves peer-to-peer technology, communication protocols, distributed systems and novel approaches in video compression. All these technologies are under development. Metadata enhances the process in many ways: 3D graphic models are currently described in metadata like languages (e.g. VRML); metadata helps to search for digital communities with similar interests; it also helps to exchange digital content between people; it describes personality types and consumer characteristics; and helps to increase the awareness of people in connection with digital worlds.

Simple application scenarios range from simple chat systems, e-mail applications, instant messaging and teleconferencing to more advanced and sophisticated solutions. They include playing on-stage computer games which are broadcast live. The consumer is directly involved in the broadcast show, letting user communities compile their content, and letting consumer-created assets act as media productions to be broadcast.

6.3.3 Digital TV is "Alive"

Technology that enables intelligent environments where equipment observes itself, its multimedia assets and its surroundings, is slowly reaching consumer homes. Digital TV equipment is alive, suggests ideas and observes the consumer's mood and emotions to provide content at the right time to the right consumer in the right way. Multimedia assets should be delivered to the consumer implicitly according to his desires and needs. The consumer should not bother about explicitly searching within myriad megabytes for the content he desires. What about a new virtual friend? Digital TV equipment that chats and talks to the lonely exists (see Fig. 6.7 as example sketch). The aim of service creators is to develop intelligent digital TV services, which personalize, talk, support, observe and do simple tasks for the consumer.

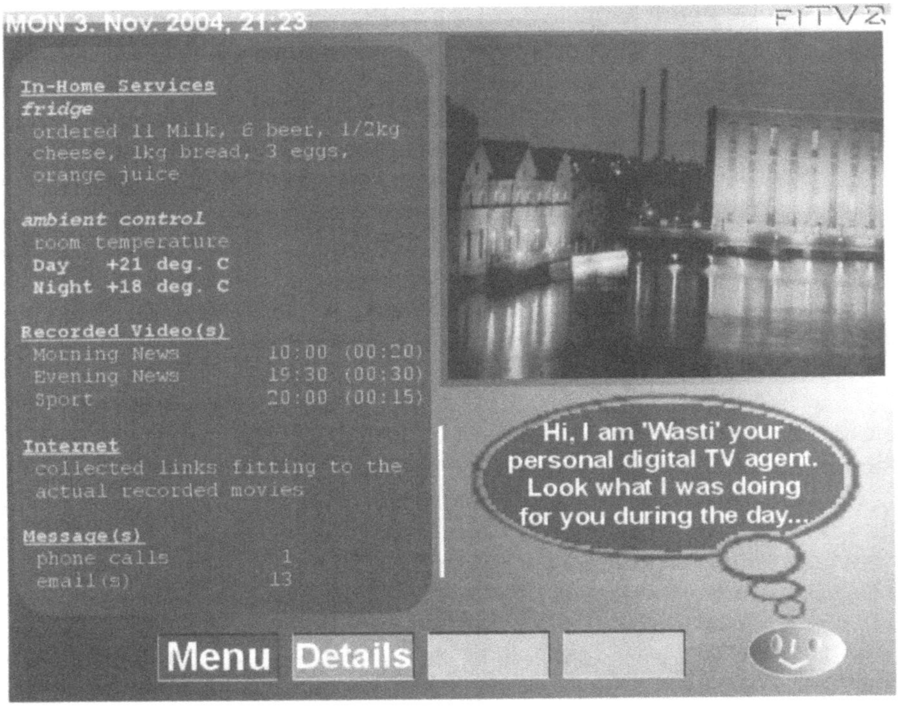

Fig. 6.7. Sketch for digital TV equipment acting as virtual "friend" — controlling the fridge, giving orders to the shop, recording broadcast shows, controlling the living room...

Intelligent environments and artificial intelligence have been a research topic for a long time. It will a while time until this vision is realized. On the other hand, technologies providing one small step farther in this direction are available. Personalization techniques to support the consumer are already available on multimedia devices. These are presented in further detail in this chapter. Agent-based technology, where small computer programs act autonomously on environmental inputs are already commonly used too.

Agents observe the customer behavior and transmit it to the broadcaster for updating consumer profiles. Metadata lays the foundation of intelligent equipment. Its descriptive character enables content matching methods for personalization, enables interoperability between devices for communication or asset exchanges and provides excellent computer-readable data models. The development of community-oriented intelligent environments is essential. The fact is that we do not watch TV solely and alone.

Example 6.7 (the digital TV as "friend"). TV equipment also orders food, collects newspaper articles, downloads multimedia assets and observes user behavior or just simply switches off the TV if the consumer falls asleep during a boring broadcast show. It also talks and communicates with its users.

6.3.4 Surrounded by Ambient "Things"

What do we mean by "ambient"? It stands for "[...] relating to the immediate surroundings [...and] creating a relaxing atmosphere" or ambiental, "[...] which refers to the surroundings of something" [53].

To include ambient services means:

- to either create intelligent systems embedding TV or TV embedding intelligent systems;
- to surround the person with sensors obtaining input parameters for the creation of service spaces;
- to make TV and its related devices available throughout the physical world;
- to see TV either as part of ambient service spaces or as a major entity within this place.

Ambient systems include more than just higher video quality, better compression techniques and enhanced user interface designs. People are embedded in and surrounded by services with transparent software and hardware, where ever they go and in their natural environment as well, simply "ambient — wireless — immersive". Ambient multimedia includes computer graphics, pervasive computation, ubiquitous computation and "old" multimedia together with a natural service space. The concept of ambient multimedia is currently under development by the European Union [95].

Pervasive computation relates to "computers and sensors everywhere in devices, appliances, equipment in homes, workplaces and factories, and in clothing [...] with a high degree of communication among devices and sensors [...] and secure network infrastructure with a wired core and wireless adjuncts that communicates with this core" [144].

Ubiquitous computation "has its goal [...to] enhance computer use by making many computers available throughout the physical environment, but making them effectively invisible to the user [...by integrating] hardware components, network protocols, interaction substrates, applications, privacy and computational methods" [171]. Ubiquitous means "having or seeming to have the ability to be everywhere at once [...by being] omnipresent" [53].

Example 6.8. Sensors all around the home watch the user's movements and present the TV content on the related screens in the room. A person walks into the kitchen and the program she was watching in the living room is shown automatically on the kitchen TV.

6.3.5 Lifestyle and the "Things" We Move and Show

Lifestyle is to present one's personality and belongings to others. Digitalization and mediated communities unfortunately allow this in a limited manner. Personal presentation is restricted to a nickname and the exchange of things means to download some bits and bytes. Introducing the aspect of lifestyle in the digital world is rather a difficult task.

6.4 Creating a Narrative Taxonomy

Realizing the paradigm shift in broadcast multimedia is essential for understanding novel approaches towards service development. Let us start by formulating a few questions.

- In which and with which multimodal environment does the consumer interact?
- How is it possible to compile several available multimedia facilities to a useful whole?
- Which space (either virtual or real) do producers provide to customers with which to interact?
- What does it mean to compile multimedia assets into a useful whole for the consumer?
- Why do consumers enjoy leisure multimedia content and want to interact with it?
- What do we understand by "interacting with content"?
- How can content be created by involving several aspects of interaction facilities, different multimedia asset types and narrative models?
- Why do people want to be told stories or entertained with narratives, especially in analog TV?
- How is it possible to create clean and complex data models enabling narrative systems?

On a very abstract level, broadcast multimedia is a story-telling environment that tells fictional or non fictional narratives to entertain or educate consumers. The story-telling medium is the TV equipment showing sequences of aesthetically and semantically ordered multimedia assets. Therefore TV tells structured narratives of a certain topic of a fictional or non fictional "real-world" happening. Actors, breaking news, viewpoints, opinions, facts, distant places, special effects and the artistic touch of a movie director create a fictional space interpreted by the consumer [153]. An experienced and creative production team truly works hard to create and build such a fictional space.

The consumer is still a passive "couch potato" mesmerized by multimedia assets created by others while interaction is reduced to zapping channels. Some do not want to interact; they only want to consume stories, sitting at a bon-fire listening to a story-teller, reading books, looking at drawings, or watching dances. Some will say that those media consumers were not passive and they are right: people could ask questions, imagine the place of happening by themselves or participate in dances. The current multimedia landscape shows many similarities. On the Internet people create their flow of stories by clicking through myriad hyperlinks, in computer games they are actively involved in how action is created and interactive computer-assisted education is all-present.

6.4.1 Content of Digital TV Is a Digital Narration

Let us take a look at digital TV with its sophisticated facilities such as feedback channel networks, advanced interaction devices and display types. Which new world and essence does the medium provide for the user? TV is a medium for narratives, which are told either by actively involving user interaction or in a "couch potato" mode. TV and its content create a fictional universe for the consumer. To go deeper and find more sophisticated explanations and formal descriptions let us review papers and articles on the subject.

To build a fictional universe for digital TV, i.e. content multimedia asset creation, means understanding happenings of the factual world and the skill of building fictive worlds (preconfiguration). The factual world is the real world (e.g. happening in reality shown later during a news broadcast), whereas a fictive world is created as would be a real world (as done in science- fiction movies). Narrative competence (being e.g. the responsibility of the movie director) is the skill of building an interactive TV narrative out of fictive or factual worlds (configuration). This means nothing more than assembling a narrative out of created multimedia assets. An example is editing news material taken from real-world happenings. This new fictional universe is presented to the consumer via digital TV. The interpretation of this narrative space is left to the imagination and fantasy (reconfiguration) of the user. This idea is based on Paul Ricoer's works for textual media narratives [153]. We extended his ideas to the domain of digital TV (see Fig. 6.8).

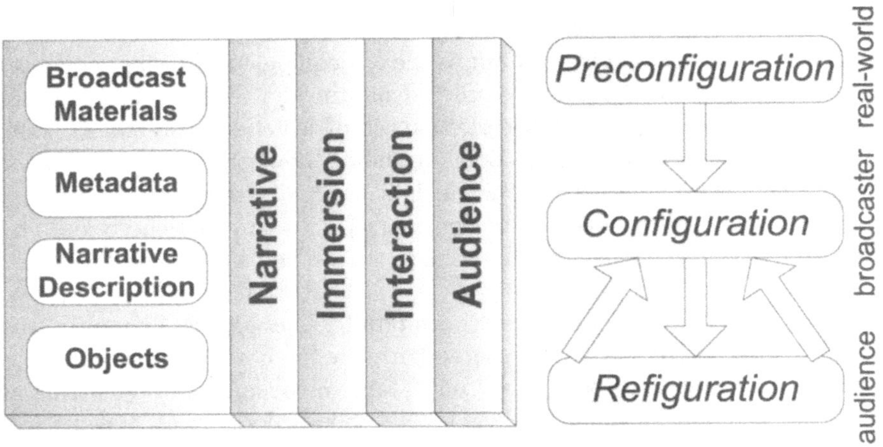

Fig. 6.8. Building a fictional universe for digital TV

Adding interactivity means guiding the consumer through multimedia assets simply by adding certain semiotics to the narrative flow. The underlying principle of creating the theory has been excellently clarified by Marie-Laure Ryan for two features in [155, 154, 156]: the first feature is the *world of fiction* and the second feature is the *development of the narrative.*

The world of fiction creates the fictional universe and sets implications for building a narrative space out of the fictive and factual worlds. It certainly sets the border for narratives. This is devoted to the overall creation of a fictional universe. The fictional universe is the place where the story flow takes place. Traversing through the fictional universe is equivalent to stepping through an infinite number of narrative states. Each state represents an atomic unit of story or semiotic evolution.

6.4.2 Creating a Narrative Model for Digital TV

Our vision for digital TV is to create a new fictional TV universe for the consumer with room for involving the consumer's illusion, interaction and fantasy. Digital TV technology enables a shift of viewpoint from the reconfiguration to the configuration state by invoking interaction. The responsibility of producers is to create complex universes in the form of compilations of video, audio, semantic structures, services, applications, interaction models and many others.[3]

[3] The narrative cube has been developed by the authors in [124] and is partly paraphrased in this work. Several enhancements have been incorporated and new aspects integrated. This ranges from more types of multimedia assets and to a deeper discussion about the relations of different spaces.

Producers have to understand this shift of paradigms and that it is necessary to provide a possibility to involve consumer interactions into the way content creators present their content. However, consumers will be able to create their own fictive universe either in passive or active positions. Passive means "couch potato"-like; active means interacting and controlling the narrative flow.

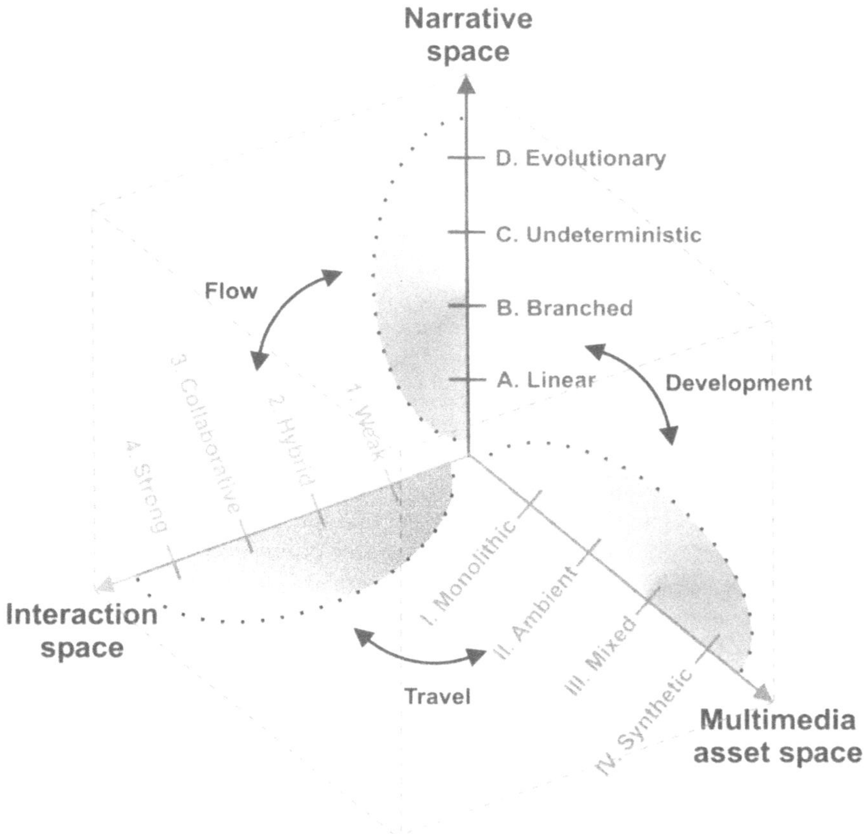

Fig. 6.9. Fictive digital TV universe — the narrative cube

A fictional universe in digital TV can be represented as a cube called the *fictional digital TV universe*. Each axis of this cube spans its own space. There are three subspaces where consumers can move freely (see Fig. 6.9): *multimedia asset space* as a place where the narrative manifests as a sense-able structure to the consumer; *interactive space* for moving the consumer from reconfiguration to the configuration stage; and *narrative space* for the development of the story and its boundaries. Multimedia assets are representations of a factual or fictive world and compiled into a semiotic content space.

Ability to navigate or travel through the content space is enabled via the interaction space. It acts as an interface and provides methods for feedback in any arbitrary form. In other words, the interaction space means relating consumer input with the presented content space within the boundaries of the narrative. The development of the narrative as such is part of the narrative space and it is up to the creator to decide which boundaries and how many story states he wants to enable.

6.4.3 Multimedia Asset Space

This axis of the fictive digital TV universe describes the relations among the consumer, the manifestation of interaction devices and multimedia assets. Embedment of the consumer into multimedia assets, i.e. immersion into a fictional universe, is a major goal for the creation of a fictive digital TV universe.

This universe includes not only simple content types, it also includes whole services, input devices and output devices as a multimodal space for deploying advanced service types. The purpose of the axis is an intelligent and semantic interrelation between several services. In short, the axis addresses consumer perception of multimedia assets and interaction devices. The degree of immersion into multimedia assets and their presentation is relevant for finding an objective measure. This includes the degree of embedment of the consumer into multimedia assets into real worlds, service spaces or synthetically created worlds. Another objective measure is the degree of relation between real-world and synthetic, artificially created environments.

This is enabled through simple presentations of multimedia assets (*monolithic multimedia assets*), embedding media as part of the real world (*ambient multimedia assets*), mixing fictional and factual worlds (*mixed multimedia assets*) and providing synthetic worlds including virtual interaction devices (*synthetic multimedia assets*).

Monolithic multimedia assets are presented to the user with very limited interaction or narrative possibilities. Passive multimedia asset consumption is predominant, as well as separate interaction devices. Presentation media and interaction devices are part of the "real" world. Subcategories of monolithic multimedia assets are *static assets*, *synchronized assets*, *adapted assets*, *real-time assets* and *evolving assets*. Most currently existing services belong to static assets, which are simple monolithic structures such as A/V streams or simple service types (e.g. EPG).

Add-on content fitting to the current stream includes various triggered events in the form of multimedia asset enhancements. Examples are advertisement banner pop-ups, content inserts, additional sound effects, additional information fitting to the current broadcast stream, music smoothers for scenes, etc. Adapted assets relate to the ability to dynamically customize multimedia assets in specific environments. This also relates to transcoding content assets or transforming multimedia assets. The high diversity of consumer terminals and available network bandwidth implicate solutions for delivering services. The simplest examples are video broadcasts adapted to the available bandwidth from the distribution channel. Real-time adapted assets alter the content itself, thus adding value to content by altering its characteristics, perception or content. Consumers themselves are able to alter content and to deploy it in broadcast shows. Evolving assets address the fact that the broadcast is in the role of providing an A/V framework, where consumers add their own creations and objects.

Ambient multimedia assets surround the consumer. Devices and consumers are also part of the "real world". Ubiquitous or pervasive devices enclose the consumer in multimedia presentation and interaction. The consumer is surrounded by intelligent technology without being aware of it. Intelligent, small-sized and distributed computers enable such environments. The interaction device hardware is disappearing and the presentation equipment and multimedia assets converge towards a unified service space. Thus, the invisible devices surrounding the consumer are becoming part of a set of interaction devices. Ambient also means adding intelligence. Therefore programs can observe user behavior, gestures or movements to personalize content according to consumer-specific needs.

Developing services in the form of ambient multimedia assets integrate the natural environment, intelligent technology, disappearing hardware and software, as well as create an ambient culture for digital communities. Ambient technology acts as an interface among the user, technology and the embedded environment. It enables the development of services including aspects of enjoying content anytime, anywhere and anyhow. Examples are services including speech- or vision-based interfaces, sensors observing user behavior and acting directly on the presented broadcast show or controlling technology. While viewing a TV show the consumer e.g. gets into his car and drives away from his home leaving the TV on.

Example 6.9 (moving location). When viewing a TV show, the consumer disappears into his car and drives away from his home. An intelligent living-room computer recognizes the departure and forwards the TV show to his car-PC.

Mixed multimedia assets embed the multimedia asset into the real world and merge elements from fictive and factual worlds. Thus content becomes transparent and merges with the real environment. Where ambient multimedia assets embed devices into the real world, mixed multimedia assets embed content elements into the real world. The consumer experiences a mixture between fiction and reality. Mixed multimedia assets do not replace the real world, they only enhance the way it is presented. From a movie production point of view it means adding special effects, manipulating content in real-time and creating a fascinating mix from real and fictive content.

The whole mix is perceived by the consumer as one reality. Other examples are mixed and augmented reality, where computer graphics are inserted into real world video streams in real-time. The consumer can take a walk through a town with a special display mounted in front of his eyes acting as an advanced mobile screen. During the walk a computer system inserts graphics as real objects into the actual world he is seeing. Teleconferencing and tele presence are also newly emerging. Specifically for digital TV, a fascinating feature can be the viewing of movies from different viewpoints and perspectives. A consumer can e.g. choose to watch the movie as if he himself played the lead role. Narrative and content will adapt to this desire. The viewpoint of the story flow is based on his personality and he himself is the lead actor. Figure 6.10 shows mixed multimedia assets during the production of a TV show.

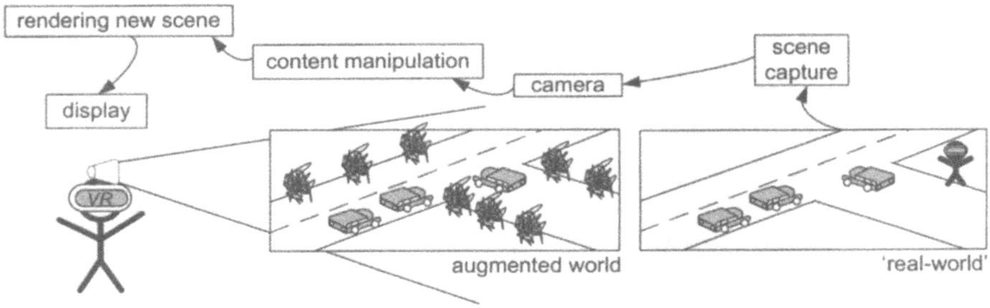

Fig. 6.10. Use-scenario for mixed multimedia assets in movie production

Synthetic multimedia assets mean a temporary "real-world" for the user. The consumer gets absorbed into a synthetic world and interaction is only enabled within the synthetic worlds. Interaction devices are physically present in the real world, but the user is only aware of them through the synthetic world's presentation. Complete temporary de capsulation from the real world and getting immersed into a new synthetic world requires artificial stimulation of human senses. The highest degree of synthetic multimedia assets currently available are virtual reality and computer graphics. Persons are embedded into a computer-generated world by different types of equipment ranging from head-mounted-displays to CAVEs. CAVEs are small chambers, where 3D graphics are displayed on the walls. With special glasses the user is transmitted into a computer-generated world.

6.4.4 Interaction Space

Interaction is the ability and capacity to alter the way multimedia assets are presented at the consumer side. Interaction either influences the narrative or lets consumers travel through multimedia assets. It ranges from simple navigation through assets, up to traveling through complex synthetic worlds. In-

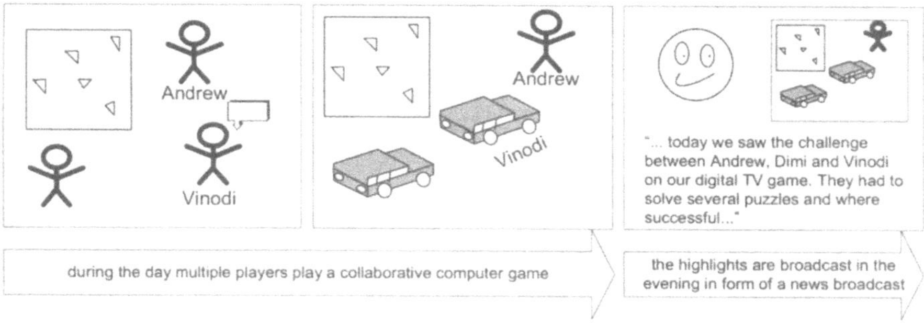

Fig. 6.11. Use-scenario for a synthetic multimedia asset: computer game as TV show

teraction is enabled through interaction technology (e.g. input devices) either explicitly or implicitly. Explicit interaction obtains user input from interaction devices present in the factual world or their presentation in a fictive world. Implicit interaction invokes sensors, actuators and intelligent devices for unconscious user interaction. In other words the interaction space relates different categories of interaction to consumers and interaction technology (e.g. potential input devices) by applying interaction models (see Fig. 6.12). To see the interaction model only as part of the interaction space is the wrong viewpoint, as it relates or connects several different axes and depends on the narrative as well as on asset types.

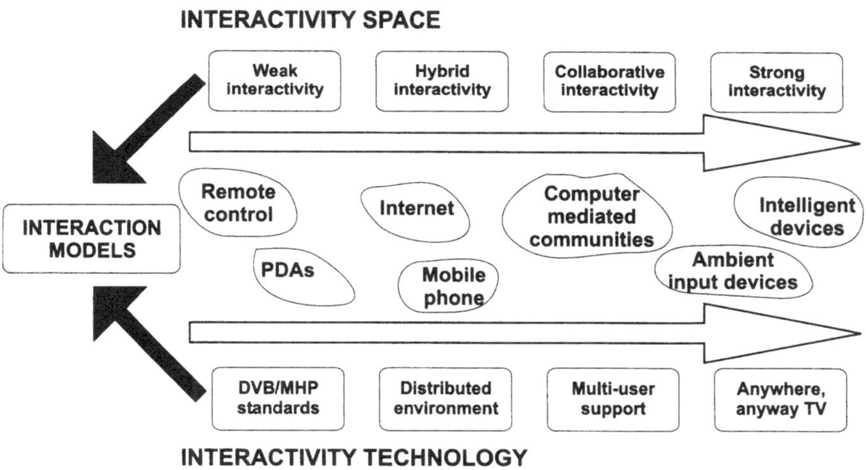

Fig. 6.12. Interactivity categories related to technology and interaction models

The interaction space is a place for gathering user interaction either from single consumers or mediated communities to alter multimedia asset presentation and narrative flow intelligently. The interaction space ranges from simple *weak interactivity* to *strong interactivity*. Weak interactivity is based on well-known interaction models. These include transactional or informational services involving typical consumer devices such as remote controls, keyboards or mice. *Hybrid interactivity* adds distributed systems and network technology such as interconnected PDAs, mobile phones, virtual reality equipment, touch less mouse pointers or Web-camera-based input modes. *Collaborative interactivity* adds computer-mediated communities and community behavior models into the form of how interaction is performed. *Strong interactivity* enables certain device intelligence based on the principle of sensors and actuators involving a consumer's or community's unintended interactions.

In general, human-centered interactivity requires [90]:

- adaptive — responding to user inputs in a context sensitive manner,
- reactive — reasonable and unprompted responses to events-based on user behavior and
- empower — automatic extension of system functionality based on user interactions.

Weak interactivity is currently all-present in existing multimedia systems. Human–computer interaction models are well researched. The consumer has to learn new interfaces and technology characteristics. Simple interaction technology, such as keyboard, remote control, joystick, mouse or pointer device is dominant. Digital TV already makes use of those and a more detailed description at this point is not required.

Hybrid interactivity adds the distributed networked technology to interaction facilities. Weak interactivity is mostly related to human–device interactive technology. Hybrid interactivity enables more advanced interaction forms. It interconnects any type of interaction device. Network protocols, lightweight operating systems and advanced interaction models are present. The equipment involved consists of PDAs, mobile phones, virtual reality equipment, Web-cam-based input devices, voice input, laser-pointer devices or other gadgets including their wired or wireless communication protocols (e.g. Bluetooth®, WLAN, LAN). Figure 6.13 presents a broadcast show displayed on a TV set. Consumers can enjoy parallel additional information on a PDA. Such forms of interaction are required for not disturbing others when they are enjoying their TV show. During an ice hockey match the information about players could be displayed on a PDA, while the match is shown on TV.

Collaborative interactivity characteristics are inter human communication, exchange of assets, awareness of others, trust and privacy, performing collaborative actions, sharing experiences and sharing of knowledge and trust [174].

The general aim of collaborative interactivity is to provide a possibility to form digital places to gather people. Consumer interactions of a virtual or mediated community are accumulated to achieve an arbitrary team goal.

Fig. 6.13. Sketch for obtaining additional information via PDA, while the other consumers can enjoy TV shows undisturbed

Digital communities form themselves according to the same interests and goals as the real-world communities. In digital TV, services for computer-mediated communities might be chat systems, discussion forums in parallel to a running broadcast show, video-conference systems, or playing a broadcast computer game. Peer communities form on the basis of a digital society, where common interests are based on social understanding. The distance of communication and interaction is world wide and happens on the application protocol layer.

Strong interactivity is based on previous interaction categories and enriches interactivity by the integration of intelligence into technology. Consumers' or digital communities' unintended interactions or estimated interactions are involved. Technology functionality increases, but its usability decreases proportionally to functionality. More device functionality does not essentially mean easier usability or that the consumer uses the complete set of functionalities. Integrating intelligent computer-assisted interaction facilities into the gap between usability and functionality decreases and makes complete device functions available to communities or people.

Intuitive interaction models are all-present in a ubiquitous pervasive powered computer environment. Seamless integration of intelligent interaction equipment into the real world or in synthetic spaces enables computer-assisted interaction. Interaction is distributed, world wide and is embedded into the natural environment. Strong interactivity also means interaction in synthetic worlds, by presenting interaction technology synthetically (e.g. display of a virtual hand in a virtual environment or display of a virtual remote control in a synthetic world). Interactive games in a real city environment are one example in digital TV. The consumer moves through the suburbs with a mobile phone equipped with augmented reality glasses and interacts with other players through them. The show could be "Big-Brother"-like with a camera team following the different players. Based on their actions a show is broadcast every evening.

6.4.5 Narrative Space

The creation of narrative TV evolves based on the simple formula: "narrative = synthetic space + immersion + interactivity + community". The narrative space is a place for immersing people into the story flow, creating virtual communities, communicating with each other and obtaining multimedia assets in a natural and narrated form. Anything in digital TV is a story! To develop a measurement or model for a story is a difficult task. The narrative space builds a narrative evolving within certain borders and has a specific theme and participants (e.g. protagonist–antagonist). It is crucial to have a certain categorization and description to provide a technical platform for narratives. Figure 6.14 gives an overview of potential story evolution as defined by [157].[4]

A *linear narrative* is very restricted as far as story development and services are concerned. Most passive services where the user cannot alter any flow belong to this category. Branching, i.e. developing the story flow along some decision points, is enabled via *branched narratives*. The consumer or digital community can alter the flow with the help of decision points. *Undeterministic narratives* allow more complex narrative spaces and direct involvement of people in the story flow, by softening story boundaries and places where stories may take place. The highest form of digital story-telling is the *evolutionary narrative*. Communities create their own place where stories happen as well as topics, gadgets and things that get exchanged. In the following a more detailed description of each narrative type is given.

A *linear narrative* means to create a top-down, tree-like story flow, where the ending and the beginning are known beforehand. Analog television provides this type of narration by delivering one piece of non interactive content to the consumer. The consumer himself has to prefigure and visualize the overall story. There is no possibility of changing the flow. It is only possible to change channels. Services are static presentations without any interaction facility and can only be consumed (e.g. stock news-ticker on a digital TV screen).

[4] The development of the narrative space is strongly based on existing works of Marie-Laure Ryan and Jon Samsel. They provided a common framework for extension in [157, 155, 154, 156].

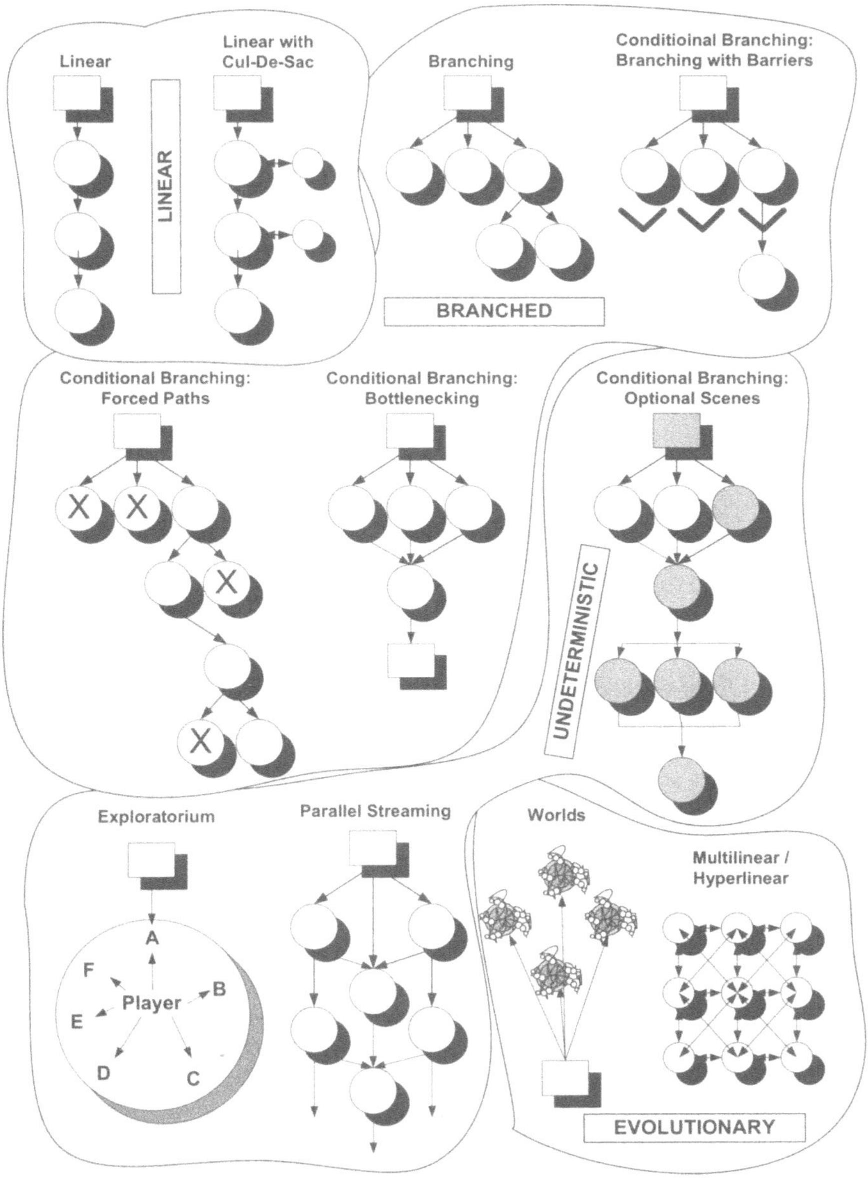

Fig. 6.14. Narrative models for interactive media [157]

Branched narrative adds branches to enable story evolution by different alternatives. Decision points are the root of the branch. Explicit or implicit choices are possible. The walk through the story path is either based on consumer input or it evolves on predefined patterns. Such patterns can be provided by personalized automatic systems or broadcaster-related inputs. Usually the starting point(s) and ending point(s) are known from the beginning.

Undeterministic narratives evolve along more complex narrative structures than simple branching. The consumer controls space, i.e. the place where the action takes place and the story evolves. They are open and nonlinear. However, a narrative director creates limitations, places, guidelines, story pieces and objects where the consumer can interact. Examples of these narratives are computer games converged with broadcast content and putting consumers live onstage into a 3D computer scene. The consumer can alter his own objects and visual representation and the outlook of the virtual scene. Another example is a cooperative broadcast game, where a digital community has to solve puzzles to gain access to the next level. Only the story background and current status are broadcast regularly to keep the consumers up-to-date.

Evolutionary narratives have the highest degree of freedom from the consumer point of view. Topical restrictions are softened and communities create their own story flow. It is impossible to predict how a story ends due to its self-evolving character. This means the ending of the story is not predictable at any point of time. A chat channel provides infrastructure for chatting, but it cannot determine about what people are chatting. Digital TV provides a facility to create computer-mediated 3D TV shows. They can be embedded into the real world or in complete computer-generated synthetic worlds. The consumer is represented via virtual avatars interacting with the happenings live onstage. This chapter shows a few more scenarios in this direction. Another scenario is virtual actors represented by intelligent computer programs that replace the existing real person.

6.4.6 Relating Spaces of the Narrative Cube

Interaction means to *travel* through multimedia assets. Clicking through hyperlinks, navigation through computer graphic-generated worlds or selecting content are examples of the relation between the interaction axis and the multimedia asset axis. Vice-versa, multimedia assets are sequentially presented to the consumer according to the choices he takes. The relation between the interaction and the narrative means the creation of a *flow*. Multimedia assets let the consumer build the fictional universe through which he is traveling. In other words, the narrative flow is established by consumer interactions. In the case of linear narratives, only one path through the narration is possible, due to the monolithic character of multimedia assets. Exceptions are stories in stories, where parallel narrations are shown as linear sequences.

Example 6.10 (Sliding Doors). The movie "Sliding Doors" is a good example for a story in a story. Two stories are shown within one movie. In one story the lead actor misses a subway which can be cached in the other story. This means one movie presents two stories shifted by a few seconds. The decision point is a little girl, which gets moved, where the lead actor catches the subway in the first story. In the second story the little girl blocks the way that the lead actor wants to take and the subway passes.

The *development* of the narrative, thus its presentation to humans, manifests through multimedia assets. Multimedia assets visualize the fictional universe and provide a place for interaction. Narratives limit interaction possibilities and multimedia asset development.

The simplest relation between the axes can be based on how a complete narrative cube can be delivered to the consumer as one entity (e.g. digital item) in push, feedback and distributed digital TV channel modes. The complete narrative cube is perceived as a complete unit by the consumer (e.g. a full broadcast show). This also includes temporal aspects. Within the next paragraphs we compare narrative models as introduced in [157] with the potentials in digital TV.[5] Another way of relating the axes is more top-down and relates different models to each other. We also present this approach as an essential one.

The *push delivery model* is the predominant delivery mode in current analog television where only linear narratives can be told. There can be parallel stories within a movie or they are distributed over two channels. Parallel streaming requires the interconnection of certain states of the broadcast. This is built in the form of story paths between them. Story paths can be either through a common topic or decision points where the consumer can change channels to view the next parts from different viewpoints.

An example of parallel streaming is when different story scenes in different places are broadcast simultaneously and the participant has the possibility to zap between them. But the consumer has a relatively limited possibility to influence the story development as he gets the content presented. The multimedia asset is one monolithic structure or eventually an ambient multimedia asset present. Weak interactivity is only present in the push delivery model. Customer polls, telephone feedback, SMS feedback and statistical consumer behavior observations belong to push delivery with weak interactivity. Linear with cul-de-sac narratives is enabled by a sequence of nodes that are delivered as isolated nonlinear assets. Scenarios in digital TV include the integration of add-on applications synchronized to content. They are triggered by specific content features appearing within a service. Another possibility is entertainment shows, where live telephone calls from consumers are inserted during commercial breaks.

Due to the lack of resources from the broadcaster side, the integration of branched narratives is highly limited, where forking during the story flow is enabled. A simple method to integrate would be to let the user choose his preferred movie endings or add dramatic roundups with personalization methods. A high degree of bottlenecking narratives is required. This means keeping the

[5] The authors discussed this topic in [123]. The comparison between the different narrative models is based on descriptions given in [157].

original structure of the broadcast stream by converging story pieces from multiple branches. Examples include movies, where the story reaches a decision point and for a while the story is told from the perspective of two actors (e.g. male/female). At a certain point in time both branches converge again. The delivery of exploratoriums enables the consumer to pause the program flow to interact with current scenes. During a scientific broadcast show the consumer would be able to pause the TV broadcast to perform chemical experiments fitting to the just-explained theory. Experiments could be performed with simple computer applications.

The *feedback channel model* allows more advanced models of interacting with assets and narratives. Several structures from push channel models are also applicable in feedback channel models. Invocation of advanced conditional branching methods by the integration of feedback from either one or multiple consumers allows collaborative interactivity as well as weak or hybrid interactivity. It is possible to add barriers and puzzles or obstacles that have to be solved either from one or multiple consumers. Puzzles have to be solved before the narrative flow can continue. A feedback channel model-enabled branched narrative would be a short quiz show between two story pieces that would have to be solved by the overall audience. Limiting the number of choices of the narrative flow leads to forced path or bottleneck structures. For example, during an interactive TV computer game, only one solution would solve the puzzle upon which the broadcast narrative continues. Another possibility for saving resources is to integrate branching with optional scenes, where consumers or consumer groups can select different alternatives. They can be in the form of a story (e.g. additional information for understanding the broadcast show completely) or certain objectives (e.g. informational multimedia information).

The *distributed model* requires more advanced scenarios to deliver an intelligently formed story piece. From the narrative space point of view world model structures are present. All types of interactivity are enabled and allow the consumer to travel through the story paths. This could be in the form of broadcast computer games, where the consumer's assets are created and integrated.

Communities can be formed freely and create their own fictional worlds to achieve a certain goal within the narrative they are building. Also more complex narrative models, such as multilinear/hypermedia models are possible. The artist who creates the game restricts its narrative development very loosely. An excellent example is a town simulation in which multiple consumers actively take part. Consumers build their own town, including several socio-economic and political aspects. They create their own teams as well as their fictional world. The broadcaster only delivers (e.g. in form of a news broadcast) the actual happenings in the town of today.

Figure 6.15 shows firstly one dynamic behavior model for the narrative cube and secondly examples for interrelating different axes to each other. Interrelating axes means the creation of a volume in the form of a pyramid, which contains possibilities and potential of services in terms of interaction, assets and narrative. Discussion about the dynamic behavior (Fig. 6.15 top-left) relates to the different axes of the narrative cube (thus multimedia asset, interaction and metadata) to the factor time. Trajectories enable movement among different possibilities and potentials in relation to the factor time. With the volume of the pyramid as a metric for describing potentials and possibilities of the narrative cube, temporal aspects can be seen as a network of temporal relations and possibilities to interchange between trajectories of the narrative axis. One simple example for trajectories is the topical changes or stories told in other stories. Jumping between trajectories can be either caused by changes in one aspect within the narrative axis (e.g. interaction models change from remote control to PDA) or between different axes (e.g. user interaction changes the way a narrative evolves).

Let us consider the volume of the pyramid enclosed by subparts of the narrative cube. The volume describes possibilities and potentials for creating whole services. The volume serves therefore as a metric for the relative dimension from classical TV to analog TV. The volume also serves as a metric for the relative distance from no interactivity, no narrative and no multimedia assets towards their diverse potentials.

- In classic cinema the pyramid collapses to a triangle: monolithic multimedia asset, no interactivity and linear narrative.
- Classic TV or purely passive services are a pyramid around the origin of the narrative cube: monolithic multimedia assets, weak interactivity and linear narrative.
- More possibilities are offered by digital TV: multimedia assets range throughout all possibilities, as well as interaction models and narrative models.

The minimal service in digital TV is a linear story evolution and weak interaction through a simple remote control by presenting a monolithic asset to the consumer (e.g. news broadcast). The maximum degree of freedom in terms of narrative is a pyramid with strong interactivity, synthetic assets and a complete evolutionary story evolution (e.g. advanced chat forum).

Each space of the narrative cube can be expressed either by a discrete or continuous function of possibilities in relation to the factor time (t). The development of the narrative ($\mathbf{n}$) can be expressed as a 3D vector function of time (t):

$$\mathbf{n}(t) = (A(t), I(t), N(t)). \tag{6.1}$$

$A(t)$ denotes the evolution in the multimedia asset space, $I(t)$ denotes the evolution in the interaction space and $N(t)$ denotes the evolution of the narrative space.

Example 6.11 (example for narrative cube). A linear video represents a sequential flow of still frames. In other words, it means a temporal and spatial

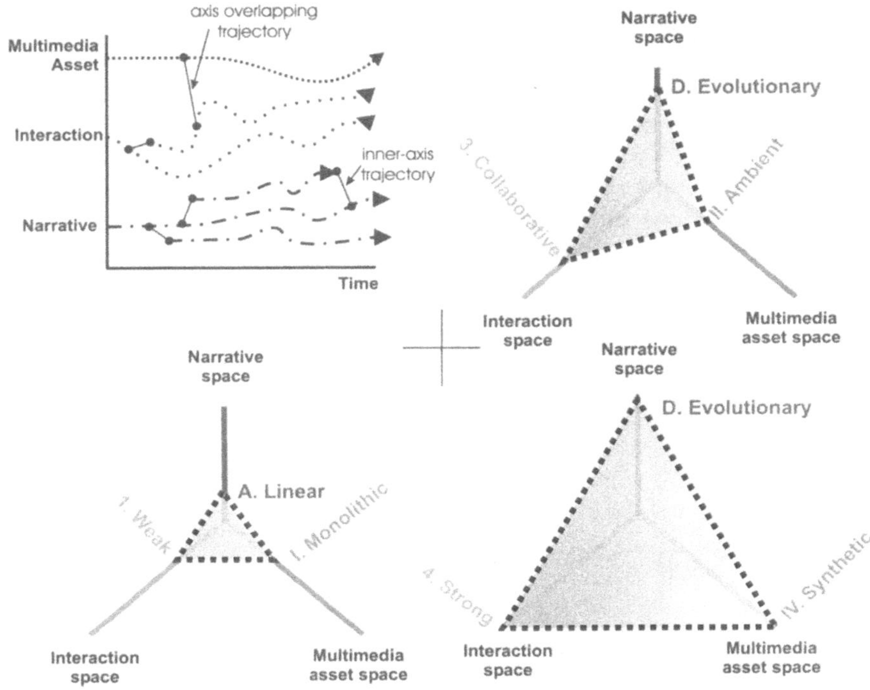

Fig. 6.15. The volume of the pyramid covering several axes of the narrative cube gives information about possibilities of each service in the spatial and temporal contexts. The factor time in the narrative cube is illustrated top left. Top right shows the narrative cube representation adapted for an ambient computer game that includes collaborative interaction and evolutionary story design as key factors. The narrative cubes in the bottom row illustrate the extreme values in digital TV and their covered volume e.g. for a MEPG (bottom left) and the maximal possibilities e.g. in a self-evolutionary computer game (bottom right).

flow of objects. These trajectories can be represented with a certain function or a set of functions. Interactivity in a computer game means to obtain values from interaction devices and relate them to computer graphic scenes. Interaction values are altering how the computer graphic scene evolves. This demonstrates the relation between two axes of the narrative cube. Both axes alter the way the narrative evolves. This is the third value of the narrative cube.

As shown in Fig. 6.15 there are two trajectories present in the narrative cube. One trajectory (inner-axis trajectory) describes the changes and potential changes in relation to time that are possible within the axis space. An example for the multimedia asset space is an actor who moves from point a to point b. His spatial movement can be described by a function. But the actor also has many other possibilities if we are looking at a computer game. There is much potential spatial relocation, therefore many more possible trajectories. The second trajectory (axes-overlapping trajectory) describes relations between axes. Let us consider a computer game as an example. The user permanently alters scenes, objects and the game flow. The relation between user interactions and how the multimedia assets change in time is expressed via axes overlapping trajectories.

Another approach is to describe the narrative cube from a more artistic viewpoint. The *actant model* helps to visualize this approach.[6] The model illustrated in Fig. 6.16 sets each narrative universe, i.e. the content of the narrative, which consists of sender, receiver, helper and antagonists.

Delivering the narrative cube to the consumer means putting the actant model into the context of technology development. The broadcaster and its service provider act as creator and deliverer (sender) of the narrative cube (object) desired by the consumers (receiver). Technical limitations and inabilities hinder the development of advanced and more immersive story environments. Therefore technology, standards and limitations act as antagonists. Innovative ideas in technology development or creative multimedia assets help to overcome the antagonists. The consumer sees the narrative cube as an object, while he is the receiver of a more immersive experience.

The vision is to create computer-mediated narrative environments, where the actual content is created using the actant model. Normally the actant model describes narratives as such. So far we described the actant model in a technology context and how technological achievements can be described with this model. Similar constellations are all-present in story-telling. The subject desires to obtain a certain object. This can be a princess or the story goal. The hero is supported by helpers, but hindered by antagonists.

Figure 6.17 shows another view of the dynamic behavior of the narrative cube. It relates several aspects of the three different axes into the context of a more reality focused model of presenting an interactive narrative. *Configuration* addresses the tasks interactive content authors have to do. It includes creating the synthetic world, implementing scripts and writing the story among others. *Backstage* procedures provide the consumer with the possibility to create his own arrangements. The consumer could change the character of actors, scenes, avatars or simply select story flows he desires. *Onstage* addresses the presentation of the narrative by involving consumer interactions. The extent of involved interactions depends on the actual configurations as provided by the author of the narrative. The difference between this and the other presented models depend on the integration of backstage procedures, onstage procedures and presentation.

---

[6] Unfortunately we were not able to find out who the source for this model is. This model was explained to us in discussions with colleagues during conferences. An Internet search pointed to [94] as a potential source for the actant model. Due to its value we added it to describe the narrative cube comprehensively.

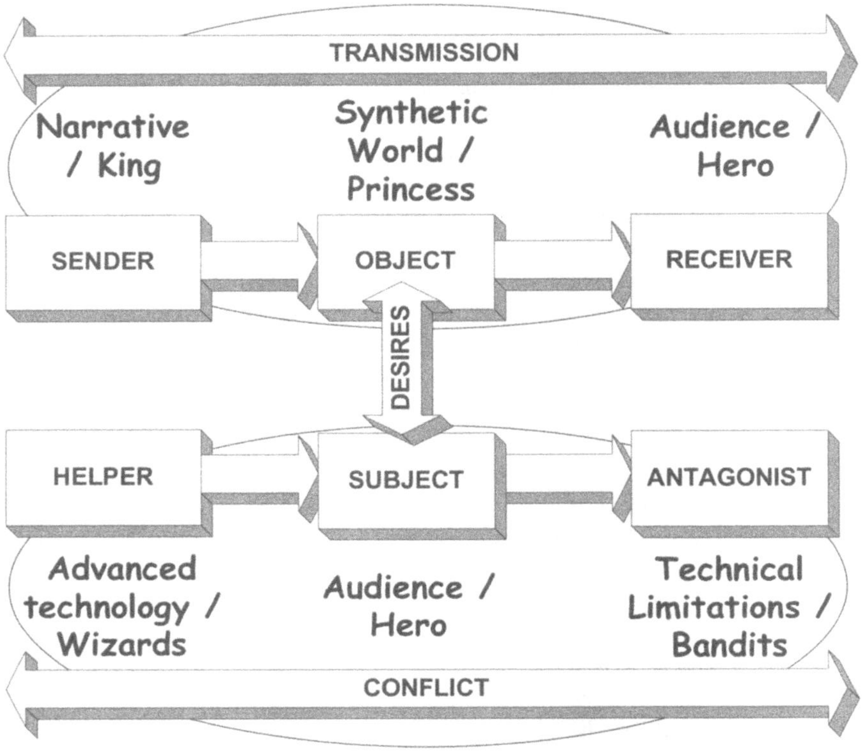

Fig. 6.16. Actant model [94]

6.5 Consumer Model and Narrative Cube in Metadata Context

Some models of consumer viewpoints and the narrative cube were described in this section. But what do they have to do with metadata? How do they relate to each other? What does MPEG-21 have to do with those topics? The answer is simple: DIs constitute a representation of the overall narrative cube. They unify several aspects of the interaction space, the narrative space and the asset space. Therefore the narrative cube unifies several concepts in an intelligent and useful meaning behind the pure descriptor definitions.

The consumer model (see Fig. 6.4) relates on a very high layer to metadata. The consumer model provides some guidelines and reference points for consumer-centered services. It influences the way metadata-based services are created. Seeing it top-down, it means to design innovative services according to those guidelines, rather than developing specific metadata models for new

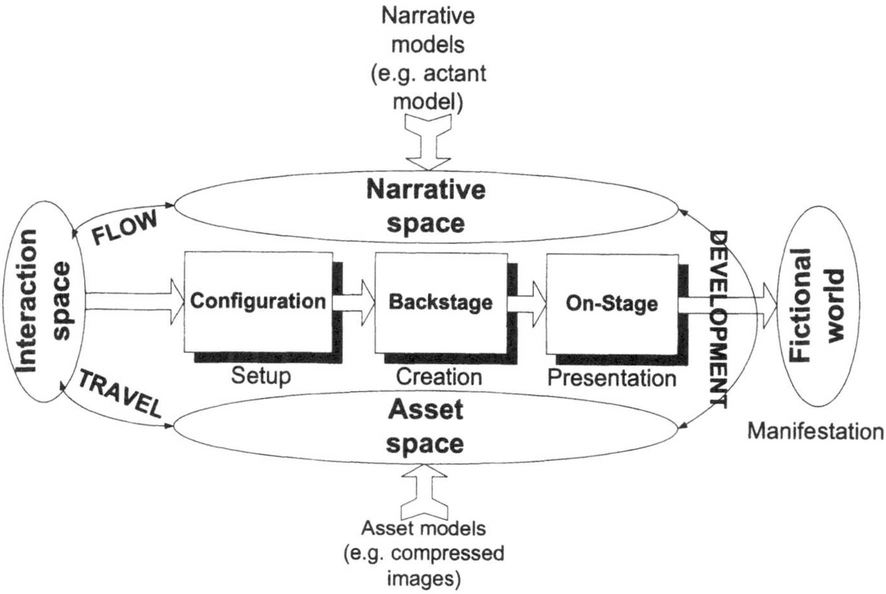

Fig. 6.17. Dynamic behavior of the narrative cube

aspects. However, viewing the design of innovative services in digital TV only from the technological viewpoint is wrong. The consumer model helps to get an idea of what consumers might want.

In a metadata context the consumer model and narrative cube mean:

- packaging several components of the narrative cube in one entity;
- bounding and limiting the possibilities of how the narrative can be developed;
- flexible usage of any type of metadata language;
- provision of a unified model for packaging narrative universes;
- integration of consumer-centered aspects into service design;
- top-down approach for reconfigurable services;
- creation of libraries and templates for reusing broadcast narratives;
- creation of abstract models for describing interaction, technology, narrative and consumer behavior;
- reconfigurable and nondeterministic stories;
- top-down approach considering technology, psychology, community and sociological aspects.

There are many models of how metadata enhances the process of delivering the narrative cube. Metadata enhances the narrative cube in many ways. Advantages of metadata range from delivery models, unified exchange and enhanced interoperability towards intelligent content descriptions. As the narrative cube unifies several presented models under artistic viewpoints it acts similarly to 3D world descriptions. A 3D model of a computer graphic model describes a virtual scene. The 3D model itself is nowadays represented with a textual description of how the virtual world is built (e.g. VRML). The narrative cube does it in a similar way, but is more complex. The unification of several aspects of delivering advanced services to the consumer is achieved. Metadata models are arbitrary and they depend on the context in which the narrative cube is applied.

Example 6.12 (consumer as a hero). To make a narrative develop as if the consumer were the lead actor is an application scenario for putting the narrative cube into a metadata context. The broadcaster delivers the narrative cube as a construct or framework of how the narrative could be told. At the consumer side, metadata descriptions stored on a multimedia home repository describe the consumer character and psychology. The broadcast narrative cube and the consumer metadata are matched and the digital TV equipment automatically builds the narrative based on the behavior of the consumer. It is also possible to alter the outlook of the lead actor and adjust it to that of the consumer.

7

Conditional Access, Digital Rights Management and Security

A very important and essential component of digital television remains to be discussed i.e. *protection management (PM)* for content and transactions and *intellectual property rights (IPRs)*. PM is a supportive technology for advanced services. Table 7.1 gives an overview of how this supportive service type relates to our key concepts.

The digital age has brought a huge change for the digital content production industry. The music industry for example clearly oversaw the Internet boom and the potential new revenue models. The users discovered the potentials of making high-quality reproductions of multimedia assets and the Internet as an illegal mass distribution medium. For the movie industry a similar development is taking place today. The Internet is an excellent source for A/V material. High compression rates and the increase of available bandwidth allow economically feasible distribution of movie material. In broadcasting the issue is more complex, as anyone can record multimedia assets onto a hard disk. From there the digital content might be spread over the Internet within hours.

Solutions to manage intellectual property rights are required throughout the value-chain in the form of asset protection management. Methods, possibilities and techniques to warrant intellectual rights on services are called *protection management mechanisms (PMM)*.

The management of intellectual property rights throughout the value-chain is an issue for protection management mechanisms, which have to be seen in a bigger context. PMM collect methods, possibilities and techniques to warrant intellectual rights on services and to increase data security to prevent unauthorized access and misuse of data through the digital value-chain. This includes end-user authentication, encryption of business transactions and protection of storage media.

The goals of PMM on an abstract level are:

- warranting and management of intellectual property rights (IPRs) (e.g. copyright protection);

Table 7.1. Relating consumer conditional access, digital rights management and security to our key concepts

Narrative Cube	
interaction	focus on the protection of the feedback channel
narrative	no direct impact on the narrative
asset	support for any type of assets aimed at
Digital Broadcast Item	
metadata definitions	DRM is directly in the focus of MPEG-21, description of used encryption technology, user-rights management, rights dictionaries...
system architecture	protected channels, encrypted data, secure database storages...
dynamic behavior	real-time encryption, DRM authorizations...
local facilities	high integration of DRM system solutions, user identification technology (e.g. smart-card technology, biometrics), authorization systems, privacy systems...
communication model	focus on secure transmission technology...
asset representation	asset representation is independent of utilized encryption schema...
Consumer Model Components	
less focus on consumer models, but easy to use DRM, CA and secure systems have to be developed	
Method and Technology	
encryption mechanisms, conditional access, biometric systems, digital rights management systems...	
Examples	
conditional access, protection of multimedia assets, use rights control, copying of multimedia assets, IPR protection mechanisms, privacy of data, secure transmissions, eShopping, eBanking...	

- secure data transmission to protect information (e.g. payment services for the feedback channel);
- management of unambiguous and machine-processable access rights to digital multimedia assets (e.g. rights to play a music file only three times);
- limitation of access to multimedia for specific users or user-groups (e.g. video-on-demand streaming upon received payment);
- protection schemes for push and push/pull environments (e.g. conditional access for broadcast content and secure transactions for eShopping)'
- user identification (e.g. smart-card solution);
- encryption and decryption tools for multimedia assets/

There are different types of classification schemes for PMMs. TV-Anytime for example divides rights management and protection into functional blocks for content protection, system protection and privacy protection in combination with an encryption and decryption toolbox [83, 82]. In this book we use the following categorization for the PMM.

- *Conditional access (CA)* is the principal technology to protect content in broadcasting. CA protects content by scrambling and providing access after verification of consumer rights. Access verification for a consumer is usually performed via smart-card technology. It has been successfully applied in analog television and also in digital television [64, 33].
- *Digital rights management (DRM)* defines techniques and mechanisms for protecting content to warrant intellectual property rights. DRM systems are collections of techniques, architectures and methods to protect digital assets. MPEG-21 standards concentrate on intellectual property management by defining *intellectual property management and protection (IPMP)* tools for exchanging messages among tools, terminals and remote locations [109]. MPEG-21 also provides a standardized way to describe a machine-readable *rights expression language (REL)* [110]. The vocabulary for the REL is described in a *rights data dictionary (RDD)* [111]. The dictionary defines semantics and terms of the REL unambiguously. A working group of TV-Anytime — namely "rights management and protection" — is also working on the standardization of DRM systems in [83, 82].
- *Security* measures apply in general cases such as in transactional services like eShopping. Security addresses either "measures taken to guard against espionage or sabotage, crime, attack, or escape" or "simply freedom from danger" [131]. Security therefore includes mechanisms for system architecture protection as well as privacy protection (as defined by TV-Anytime).

Figure 7.1 gives an overview of all three mechanisms and their concrete purpose. CA prevents access to content if a consumer is not authorized. DRM goes a few steps further and protects content and services in a larger context. It includes toolsets for intellectual property rights protection throughout the complete value-chain. Security focuses on transmission, authorization and consumer identification.

All protection mechanisms can be categorized as follows:

- *level of application:* several protection mechanisms apply on the transmission level, transaction level, stream level or service level;
- *protection management mechanism:* principles, methods and technologies utilized to warrant IPR, security and consumer rights;
- *PM system architecture:* definition and creation of an abstract framework for PM architectures, the exchange of PMMs and the communication forms between systems and system entities;
- *metadata definitions:* metadata language for describing PMs and their components.

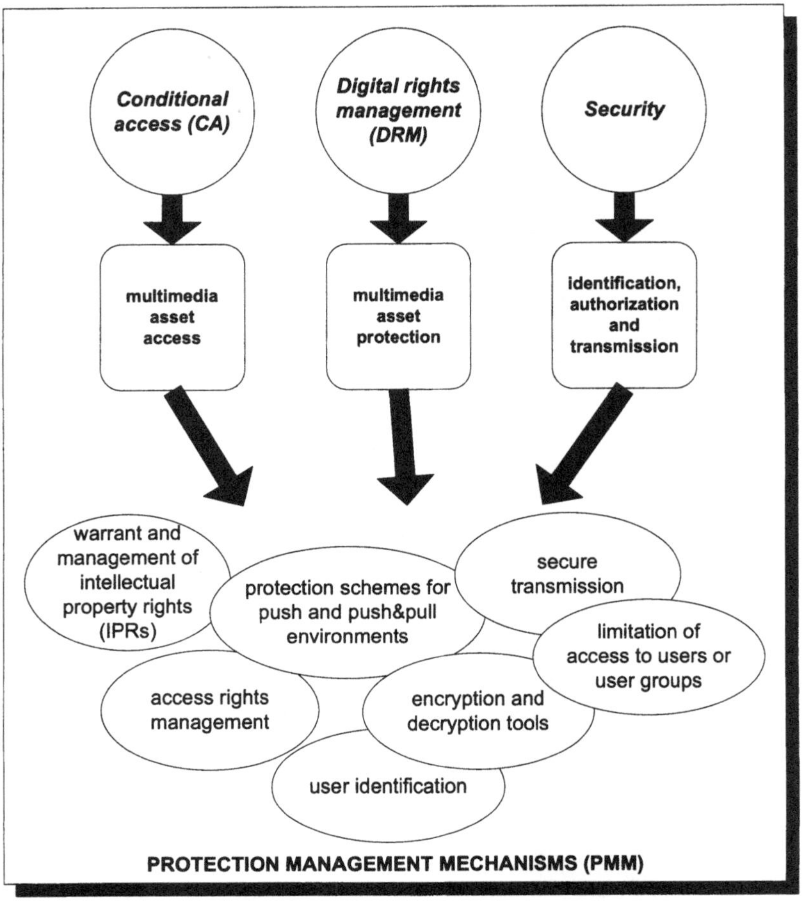

Fig. 7.1. Overview of protection management mechanisms

7.1 Conditional Access (CA)

To enable pay television it is essential to develop mechanisms to prevent unauthorized access to content. Conditional access is the most widely spread and most common solution for delivering services to those consumers who paid for them. Different solutions are available. The most common solutions are combined with smart-card technology. Consumers buy smart-cards and insert them into their digital TV equipment. They contain a key for decrypting the A/V stream encrypted with a PKI (public key infrastructure) mechanism. Another widespread solution is to rent equipment from broadcast service provider that contains the specific decryption hardware.

Europe differs a bit from other areas in the conditional access techniques point of view. A European parliament directive suggests a common interface for all consumer devices [35]. This is meant to prevent different consumer device technology implementations. Therefore DVB began to develop conformant conditional access techniques in the beginning of 1993: the *digital video broadcasting common interface (DVB-CI)* [56]. MPEG supports these efforts because of its organizational principle to leave certain components open to competitive market forces. MPEG does not define the scrambling method nor the messages that are sent to consumer terminals [45].

Let us now focus more on the technological side of conditional access. First, an encryption technology is required. The encryption has to be strong enough to prevent consumers without the encryption key from decoding the content. Second, the technology must be interoperable. And third a possibility to access TV periphery and equipment must be provided.

DVB standards allow two encryption mechanisms: *simulcrypt* and *multicrypt*. Simulcrypt enables the encryption of different streams in different ways. This means that one A/V stream is encrypted simultaneously in many different ways including its conditional access signaling. Simulcrypt also enables the interchange of parts of the programs within a network or bouquet of services. *Multicrypt* allows the running multiple conditional access modules in parallel. Several smart-card slots are installed in consumer devices capable of working with different conditional access or scrambling systems [22].

Interoperability between conditional access systems is enabled via the DVB common interface. The block diagram is shown in Fig. 7.2. The common interface utilizes a common descrambler. Its strength is its modularity, as different conditional access modules can be plugged into the receiver. Its second strength is the way communication with peripherals is performed. To prevent piracy of A/V content, it is essential to prevent external resources from accessing the decoded material. DVB solved this problem by letting the conditional access modules communicate with resources in such a way that the periphery is not aware of the details of the service performed [45].

7.2 Digital Rights Management Solutions

Complete solutions for *digital rights management systems (DRMS)* are currently under development. MPEG-21 makes great efforts in standardization. TV-Anytime focuses on more broadcast related aspects, but its DRM standards are still under development. Both standards, MPEG-21 and TV-Anytime, define their own metadata languages based on XML mostly to protect content. TV-Anytime goes a bit further and also regards security and privacy mechanisms.

The system architecture of DRMS varies between different standards. The MPEG-21 standard concentrates completely on an interoperable framework and a set of tools for managing IPRs [109]. The basic idea of MPEG-21 is the creation of abstract specifications of tools for IPR related communication between system entities and the exchange of IPR related software components and entities.

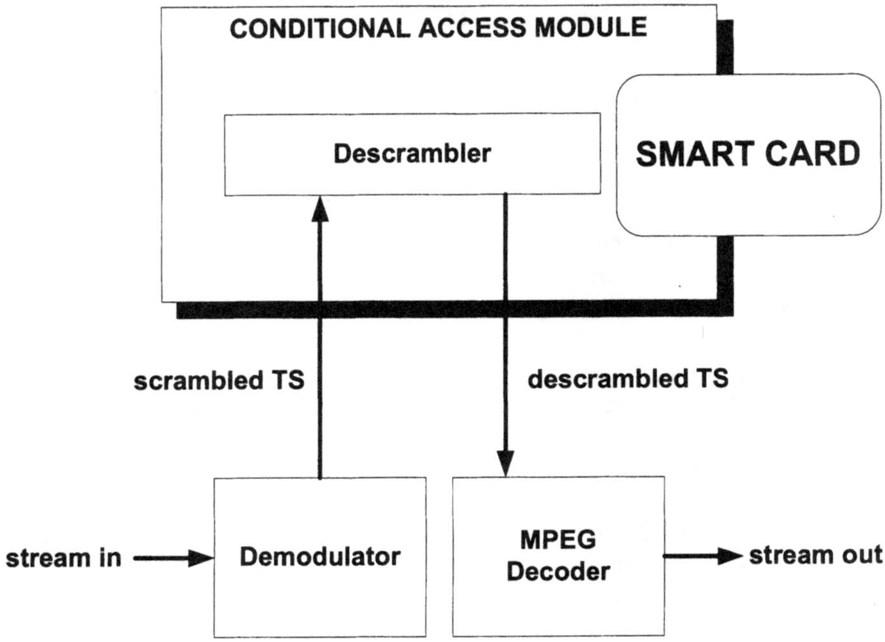

Fig. 7.2. DVB's common interface for conditional access

The content lifecycle of DRM systems has been described with an ontology [48] and is illustrated in Fig. 7.3. The creator gives certain rights to media distributors or other rights holders. Based on certain conditions they can give these rights further to other media distributors, rights holders or consumers. The rights are expressed via licenses telling which actions can be performed upon resources.

This means that a license is nothing but a collection of grants exchanged between partners. In MPEG-21 language a principal is an identified entity, which either gives or is given rights to perform actions on resources. Each exchange is giving the principal authorization to perform certain actions on a resource (e.g. copying, editing and transforming). Conditions restrict the actions that may be performed. In the MPEG-21 context this is defined by the rights expression language. REL describes the syntax of data structures for licenses and grants [49].

TV-Anytime follows a different path in standardizing DRM even though the goal is rather similar to that of MPEG-21 DRMS. The DRMS of TV-Anytime is strictly built around systems without and with a feedback channel network connection. Systems without feedback channels are standardized in phase one of TV-Anytime and mainly focus on the protection of recorded content on the consumer device. They control codes for enabling or disabling access to recorded content. Phase two of TV-Anytime adds two more use-

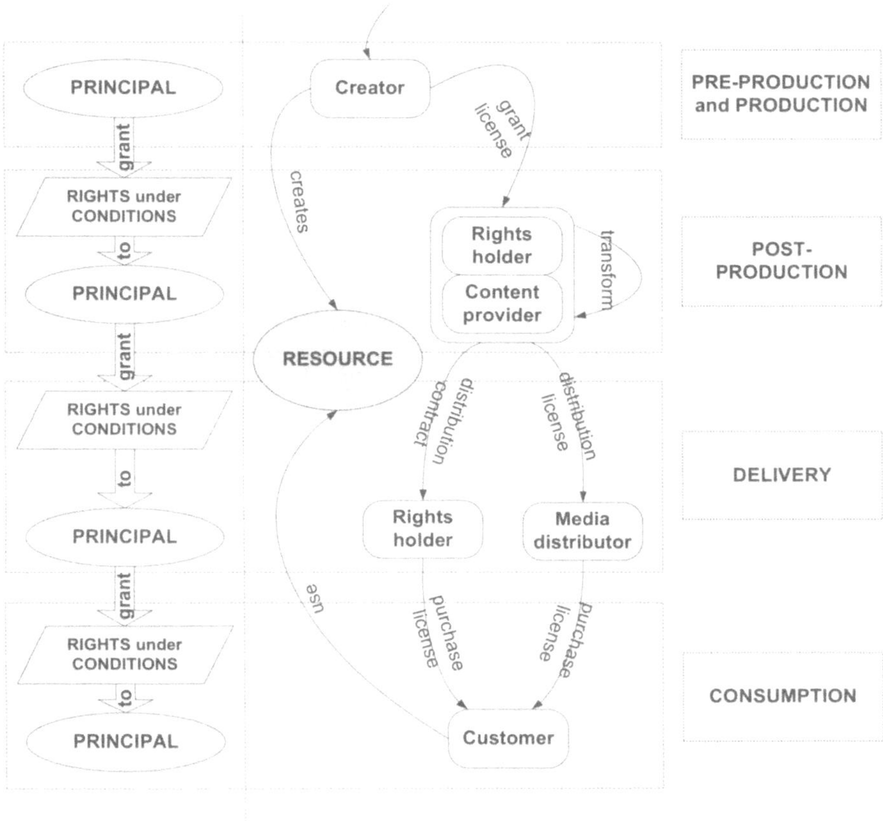

Fig. 7.3. Digital rights management from content creator to final consumer (extended from [48])

scenarios: one is for remote managing of PMMs and the other one is for real-time decryption using e.g. smart-card technology [83, 82].

The conceptual model of a TV-Anytime-compliant PM architecture consists of several application programming interfaces. They link system, operating system and hardware resources in order to disable or enable access to content. The essential parts of a TV-Anytime system are the *rights management and protection information (RMPI)* containing information about rights and conditions. The PMM itself is called a *rights management protection system (RMP)*. The RMP system is responsible for warranting consumer rights as well as intellectual property rights of the distribution side [83].

7.3 Transactions and Security

Security is used in a very wide context. For health care information systems three issues are referred to as security [141]: *data availability*, *data integrity* and *data privacy* [143].

Data integrity focuses on completeness, trustworthiness and validity of information. Specifically it means verifying user input data for certain applications. Technical systems have to guarantee data integrity in storage systems. Data models have to be concise and clean. Typically relational databases provide this mechanism. Also network protocols carrying data have to provide reliability. This addresses high-level and also low-level protocols such as IP or TCP/IP. The communication between digital receivers and other entities has to rely on protected and data-loss-free communication models.

Data availability addresses system downtimes, identification of consumers and access rights control. Reliable identification of consumers is needed to provide access to data and services to securely authenticated users. The range of technologies is rather large. Simple login and password-based solutions are extended by smart-card readers or even biometric systems. Certain data on the digital TV receiver or on the service provider side should only be available to certain consumers or consumer groups. The assignment of access rights, e.g. provided by MPEG-21, is a very essential component of secure systems.

Data privacy is relevant if systems are put into networked environments. DVB defines two separate network channels, the feedback channel and the broadcast channel. Both channels have to provide encryption facilities to warrant privacy. In some cases the broadcast channel might require encryption due to the capability to encapsulate personal data in the broadcast stream. The consumer can access e.g. his e-mail or banking information as personally encrypted data from the stream. The feedback channel typically requires encryption as it is realized over the public network infrastructure, the Internet. Clear data transmissions on the Internet cannot be used for services such as eBanking or eShopping. Internet technology offers many solutions to encrypt services. This technology can be easily migrated to the digital TV platform. To give an example, DVB supports HTTPS as the protocol for secure communication between entities.

8

Digital Production and Delivery

A suitable infrastructure for the delivery and consumption of multimedia assets has to be built and is becoming a contested issue. The role of the consumer changes to a more interactive one and requires standards fitting several components enabling interaction with content and the transfer of digital content. Due to the many metadata standards, one standard has to act as an umbrella over the others and provide encapsulation facilities. MPEG-21 aims at understanding how different components fit together in order to fit metadata standards seamlessly together. Therefore MPEG-21 seeks to create a complete structure for the management and usage of multimedia assets, including infrastructure support for commercial applications and rights management incorporated into the infrastructure.

Broadband networks enable the distribution of various new and rich types of multimedia content to the consumer. Traditional broadcasting — multiplexing only video/audio and very limited data (e.g. teletext) — is based on complex, but still not completely digitized production facilities. New digital authoring tools and methodologies are required to satisfy a complete value-chain spanning digitized production. Digitized production means not only focusing on the creation of monolithic assets; it also means focusing on the implementation of the complex interactive and enriched multimedia assets, their storage and management, packaging, versatile exchange and multichannel deployment.

Table 8.1 gives an overview of how this service type relates to our key concepts.

8.1 Wireless Protocols for Digital TV

Wireless protocol types develop along different generations providing more bandwidth, higher mobility and a seamless connection to broadband networks. Figure 8.1 gives an overview of the second, third, fourth and fifth generations of wireless protocol types and their features. For the convergence between wireless protocol types and digital TV several potential application scenarios are relevant: first, the most common use-scenario is the utilization of wireless protocols as a feedback channel protocol (e.g. restaurant advertisements are broadcasted and the consumer orders pizza with an SMS); second, wireless protocols as a carrier of TV content (e.g. TV content in cars or for the mobile, moving user over IP); and third, seamless and wireless interconnection of TV-related gadgets (e.g. Bluetooth connection between PDA and STB).

Bandwidth of wired IP networks is increasing tremendously. Wireless technology is still taking its first steps towards usable high-bit-rate connections. By the year 2010 the third generation will provide data rates of up to 2 Mbps. WLAN and other wireless LAN derivatives are already providing higher data rates and integrate seamlessly with existing wired network infrastructure. Wireless protocols underline the fact that existing broadcast cultures will change tremendously. IP-based networks, which provide data rates for digital TV distribution are emerging and render existing broadcasting infrastructures obsolete.

8.2 Broadcast On-Demand Services

Let us start with two questions: "What can be encapsulated in a broadcast stream?" and "For which services business models and payment schemes exist?". The push application environment predominant in broadcasting provides a high-bit-rate data stream. The stream is easily capable of carrying megabytes of data to the consumer home. This allows new perspectives in the development of value-added services. Technical solutions and new approaches to coping with this situation are required. Multiprotocol encapsulation, i.e. IP over MPEG-2 TS, allows all kinds of on-demand services. This scenario enables on-demand metadata-containing broadcast show accompanying descriptions or on-demand assets (e.g. video stream).

Linking on-demand services with conditional access schemes and payment schemes enables all types of revenue models. Secure transmissions and end-user authentication through smart-cards are required to obtain revenue.

Example 8.1 (metadata on demand). It is possible to stream binarized MPEG-7 data to the consumer including descriptors for available assets. The consumer can browse through them and select the services he would like to enjoy. Metadata packages containing descriptions of currently running TV shows could be bought by consumers as an additional service pack.

8.3 Capturing Metadata in Production

Capturing metadata in production is a rather difficult task. It means extracting metadata almost in real-time and delivering it exactly synchronized to

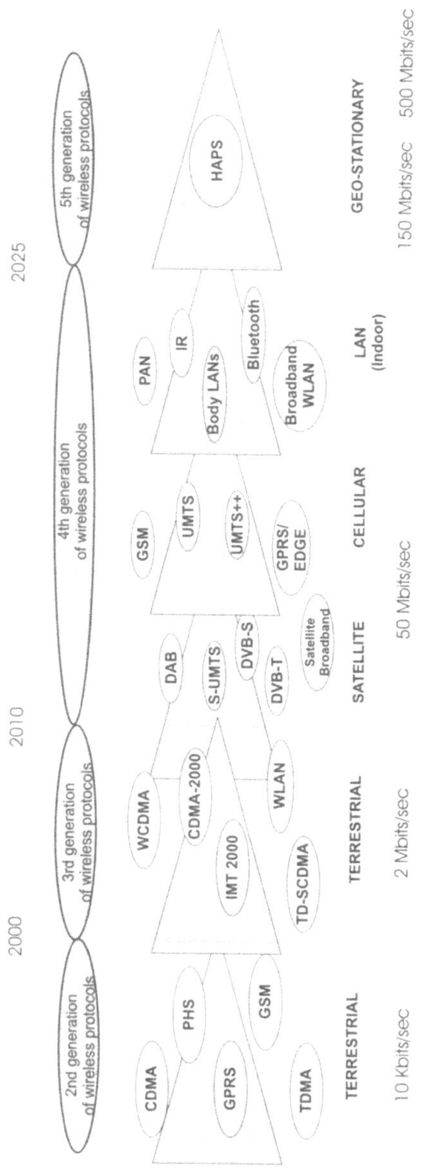

Fig. 8.1. Generations of wireless protocols

Table 8.1. Relating digital production and delivery to our key concepts

Narrative Cube	
interaction	creation of weak to strong interactivity models
narrative	narrative boundaries creation
asset	building of the synthetic space and its delivery
Digital Broadcast Item	
metadata definitions	production-related metadata definitions, capture and extraction of metadata, compilation of metadata-based services, several DBIM metadata blocks utilized, real-time constraints on metadata extraction, emphasis on basic tools and asset tools, object as well as service and narrative tool metadata are instantiated, strong involvement of vertical tools, metadata filtering, adaptation and transmission...
system architecture	professional media repositories, digital item store solutions, large-scale media production storages, typical broadcast and feedback architectures, architectures for exchanging assets between producers and in creation processes, virtual eShop solutions...
dynamic behavior	highly reliable and simple solutions in production, playout and delivery mechanisms, quality of service, instantiation of feedback and broadcast item types, focus on storage, delivery and creation phases, exchange between professional partners in the value-chain...
local facilities	creation of adaptable services, creation of interaction facilities
communication model	highly reliable data transmissions, less or no protection mechanisms in production, DRM and security for feedback services, manifold protocol types (e.g. MPEG-2 TS, FTP, RTP, RTSP, BiM, SOAP), synchronization models, real-time...
asset representation	production dependent (e.g. GXF, MXF, AAF), highly reliable and fault-tolerant solution, MPEG-2 (e.g. ES, TS), H.264 (AVS), MPEG-1, Java applications...
Consumer Model Components	
relates to creativity and ideas of service authors	
Method and Technology	
network and transmission technology, database models (e.g. RDBS, ODBS, XML), creation tools (e.g. service editing frameworks), template libraries software engineering...	
Examples	
wireless transmission of digital TV services, wireless as feedback channel network, mobile TV, wireless digital TV, broadcast on-demand services, metadata on-demand services, real-time metadata capturing, utilization of the screen-overlaying pane, real-time content manipulation, intelligent multimedia representation, video-in-video solutions, metadata-based repositories for production, digital item stores, electronic ticket exchange, travel agency information, eShopping applications, electronic postcards...	

assets to the consumer equipment. As a reference application we present regionalized advertisements, where position descriptors of advertisement areas are extracted during a live broadcast (see Fig. 8.2 and Fig. 8.3). They are delivered to the consumer terminal as well as real-time synchronized with the transmitted video. The area is covered by location-dependent advertisements. Different advertisements in different countries appear to optimize the reachability of consumers.

Fig. 8.2. Sketch: regionalizing advertisements to the requirements of different nations (Finland)

The lifecycle of capturing metadata is easily explained. Figure 8.4 shows the overall process. First, it means capturing descriptors or specific features during media production. Their storage and management in metadata archives for later use and delivery are part of this process. Transforming metadata into adequate forms for further processing in either production or postproduction is the next step. It means preserving as much data as possible and transforming them to a suitable form. Special focus is given to the synthesis of metadata: this means either using metadata definitions at the broadcaster side or consumer terminals. Synthesizing metadata means to apply the data structures that were obtained. Delivery of metadata is a more complex task and depends highly on the delivery modes chosen. Important issues are metadata multi-

Fig. 8.3. Sketch: regionalizing advertisements to the requirements of different nations (Austria)

plexing, adding timestamps and synchronization information for later usage and the monitoring of overall delivery processes.

Capturing metadata is a fourfold task. It means *extraction* from source A/V materials; obtaining additional descriptors from *sensors*; *catalysis* of existing multimedia assets; or simple automatic, manual or semi-automatic *collection* of data during production processes. Either real-time metadata capturing or offline access schemes are possible and require different performance profiles of computational power and varying algorithm development. The captured data in metadata archives including timestamps and other relevant definitions must be carefully stored.

Capturing metadata during production also means the creation of new pieces of content from the obtained metadata descriptors. This can be either in the form of multimedia databases acting as storage media or for automating tasks during production. Data are currently manually edited and reedited several times during production. Service information acting as program description is obtained from service editors, reedited and newly multiplexed into the digital TV MPEG-2 broadcast stream. Metadata accompanying the multimedia assets throughout the lifecycle eases the transformation processes tremendously. The data can be transformed in the formats required for playout and no reediting is required. The task can be performed automatically.

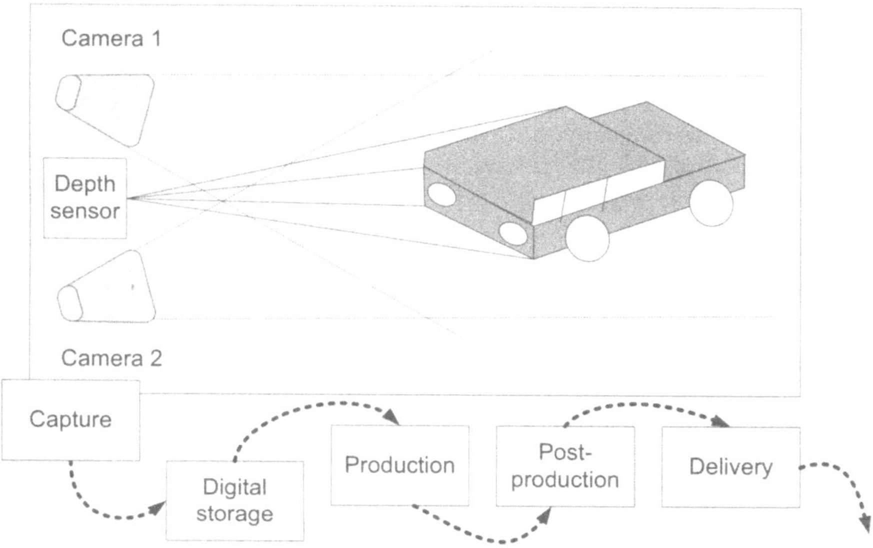

Fig. 8.4. Capturing metadata during production

Metadata extraction operates directly on existing or concurrently produced multimedia assets. It is often referred to as feature extraction or segmentation. The underlying task is rather difficult to realize and requires high computational power. MPEG left these problems to competitive market forces as defined by MPEG-7. *Sensors* capturing metadata during asset production add additional descriptors to the value of multimedia assets. Metadata that cannot be captured through custom camera devices enriches digital TV by many other value-added application scenarios. They range from adding 3D information and scene conditions, up to any other measurable parameters. The *catalysis* of metadata is performed on existing and archived assets and is not performed in real-time. It is related to obtaining descriptions from existing materials and catalyzing them to novel application areas. Metadata is mostly collected during production processes in the form of notes, documents, scene descriptions or accompanying materials. *Collection of metadata* is the easiest task to do, but requires a lot of human intervention due to nonunified data models. The DBIM especially helps and assists in this task tremendously and eases the process of unifying lifecycle steps.

Example 8.2 (intelligent multimedia presentation). There is too much information available on the Internet. Digital TV also tremendously increases the information flow to the consumer. Intelligent multimedia presentation techniques help to tailor contextual information to the consumer needs. MPEG-7 provides tools and techniques to describe multimedia assets (e.g. temporal-spatial information, encoding form of content). This representation helps and assists the synthesis in specific domains. This could include adding automated Web-links to a news broadcast fitting to the topic or to live sports broadcasts.

Example 8.3 (video-in-video solutions). Current video-in-video solutions allow viewing several channels at the same time. The user can select different channels to be viewed at the same time at specific positions onscreen. News broadcasts specifically designed for certain domains create the possibility of automating the process of viewing a wide range of content. One application scenario might be stock exchange news displayed in parallel and especially personalized from different broadcasters. Together with intelligent multimedia presentation techniques the information transmitted to the consumer can be optimized.

8.4 Designing Metadata-Based Multimedia Repositories

Metadata-based multimedia repositories have many application areas in the world of digital TV. They can be applied within the scope of any party in the value-chain of digital TV: at the broadcaster side for managing large multimedia asset archives (see Fig. 8.6); at the consumer side (see Fig. 8.5) as a home multimedia server; at the service provider side for enabling feedback channel services; at the service editor side for storing multimedia assets between their preparation steps; or for performing business logic tasks.

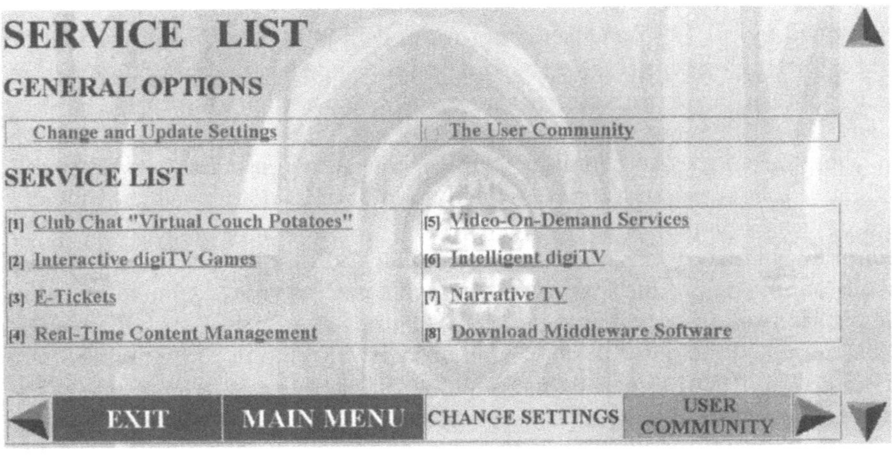

Fig. 8.5. Digital TV service portal visible to consumers

Let us describe in very abstract terms how such archives are logically designed and how their dynamic behavior is performed. Most of the components have been described in previous chapters, and Fig. 5.1 shows a schematic overview of architectural components. It is important to understand that a digital item is the atomic unit that is managed. This is independent of the architecture and of the location of the repository.

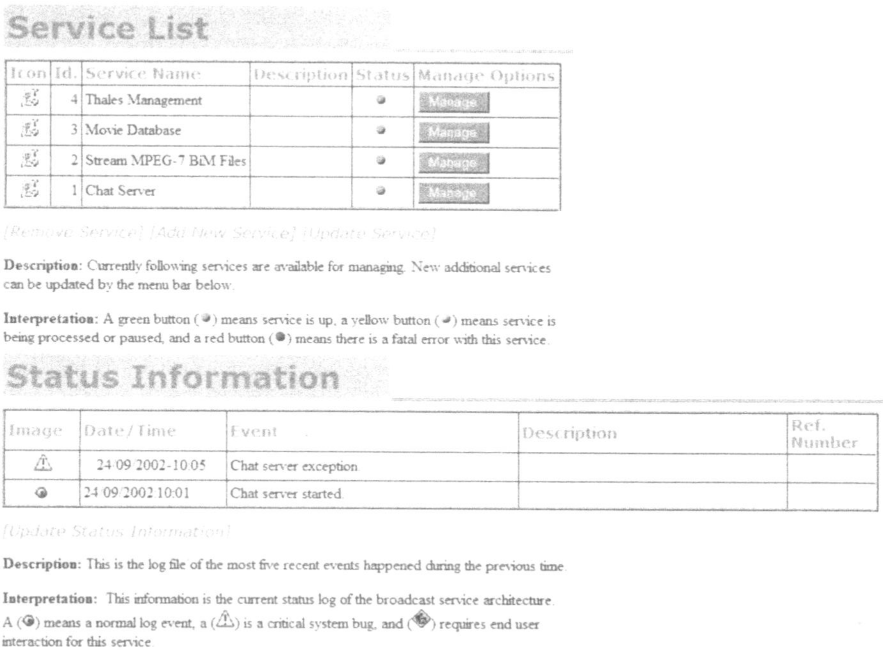

Fig. 8.6. Digital TV service portal visible to broadcasters

The application of a feedback service architecture can be arbitrary and appears in many different applications. The most common is storage and retrieval of huge broadcast video and audio databases. For playout on the Internet or within complex broadcast architectures transcoding, thus changing the format of the content in real-time, might be essential. Especially if the materials are distributed over the Internet, high-bit-rate MPEG-2 TSs are very rarely applicable due to bandwidth limitations. Therefore transcoding is essential to provide the capability to broadcast to different consumer terminals over different network resources. MPEG-21 and its adaptation mechanisms are enabling such architectures.

Metadata is the key to keep track of huge archives and the application of advanced data models enables manifold new search and filter mechanisms. But metadata must also be stored e.g. in metadata-enabled repositories to provide easy access to the metadata.

8.5 Digital Item Stores

ECommerce and selling digital goods over TV is still the major business in digital TV. Synthetic digital item stores providing digital goods in any form are one very adequate approach for eCommerce solutions. The purpose of digital item stores can be arbitrary: electronic ticket sales, ePost offices, ePostcards, eCity information services, eTax office, eGovernment office among many others. Within this section we explain the principle of digital item stores and show their application in digital TV. The specific case for electronic tickets has been presented in [128].

An integrated solution for purchasing and using digital items requires first the announcement of potential sources of digital content (e.g. during a documentary about Finland, an insert is triggered with the address of a travel agency selling trips to Helsinki). Each digital item is uniquely identified and mapped to its concrete resource presentation. The lifecycle for digital item stores is presented in Fig. 8.7. In principle, three levels of exchange are present: the B2C (business-to-consumer) layer deals with the interactions between business partners and consumers; the C2C (consumer-to-consumer) layer for exchange of digital content either between different consumers or consumer devices; and the B2B (business-to-business) layer to perform transactions between business partners.

The lifecycle of digital items in the context of digital item stores is as follows.

1. *Digital item announcement and selection (steps 1–4):* the consumer can browse through the goods delivered in a digital store and select the items he desires. Even though user identification is not required at this stage, it supports creating a personal shop appearing to the consumer. This can include shopping cards, wish-lists or personalized services.
2. *Digital item purchase and reception (step 3):* purchase mechanisms and payment schemes allow the reception of assets of different forms and over different distribution channels. Content is protected with various protection mechanisms to secure firstly the transmission itself, secondly to warrant copyright issues during consumption and thirdly to guarantee consumer privacy. User identification and authentication are essential during this lifecycle phase.
3. *Digital item exchange and usage (steps 8 and 6):* management, exchange and use of digital items depends on their type. It includes media consumption and exchange between different parties. Special protection schemes have to guarantee that only authorized end-users are capable of enjoying digital items. Portability is required to provide digital items on multiple end-consumer devices.
4. *Digital item authorization (steps 7–9):* depending on the actual purpose of digital items, authorization may be required. This is applicable especially for content protection mechanisms or digital items requiring validation

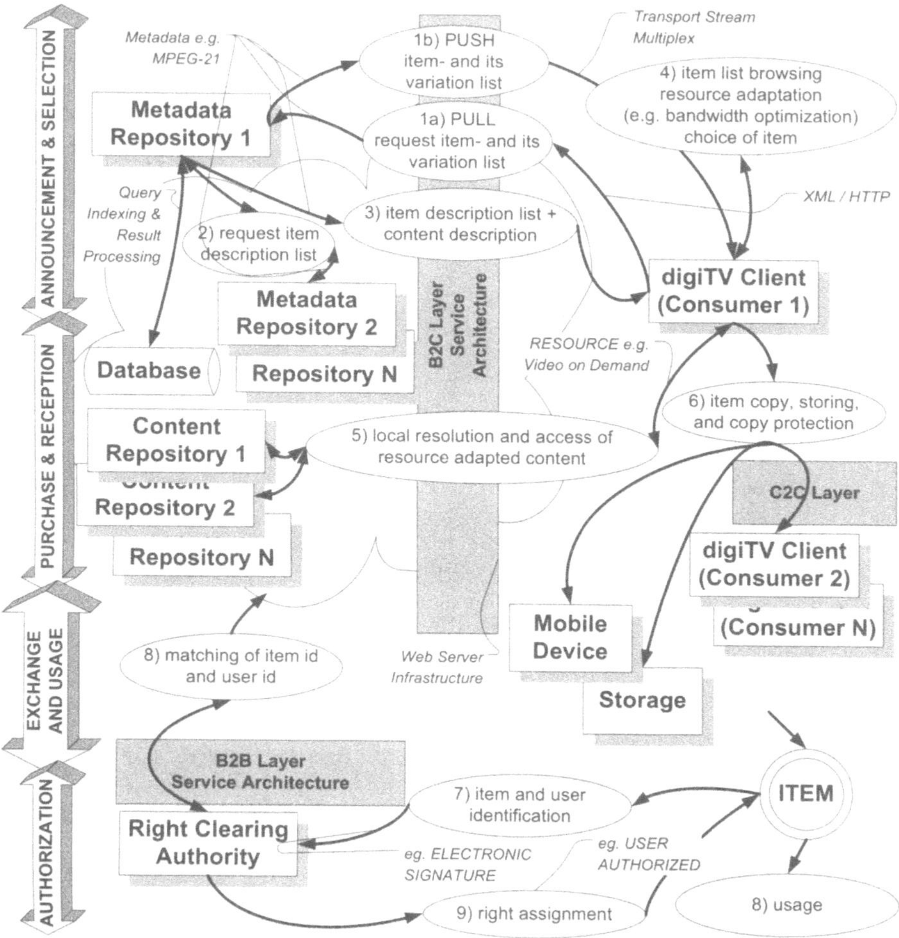

Fig. 8.7. Digital item store flow model

(e.g. electronic entrance tickets). Matching digital item ideas with user and usage rights, stamping patterns for one-time-use-only items are examples of digital item authorization. Special rights clearing authorities are responsible for performing this task.

In the following we show example scenarios for digital item stores. The first application is an electronic entrance ticket system (see Fig. 8.10), the second a virtual post office (see Fig. 8.9) and the third includes digital games (see Fig. 8.8).[1]

Fig. 8.8. Digital item store example: eGames

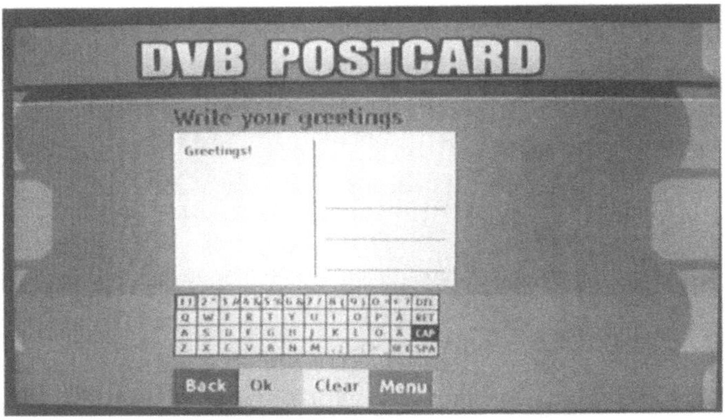

Fig. 8.9. Digital item store example: ePostoffice

[1] The applications have been developed during the Future Interaction TV Project at the Digital Media Institute, Tampere University of Technology, Finland by Mikko Oksanen, Perttu Rautavirta and Anurag Mailaparampil.

Fig. 8.10. Digital item store example: eTicket

9

Intelligently Presenting and Interacting with Content

Intelligent presentation and interaction with content is a rather manifold task. Many different application and service scenarios emerge. The list of potential scenarios is large. Some examples are stated here:

- hyperlinked (or segmented TV),
- music application,
- content add-ons such as regionalized advertisement banners,
- broadcast show enhancing sound/audio effects,
- advanced content adaptation mechanisms,
- movie scene retrieval,
- narrative and semantic television intelligently personalized to consumer needs,
- active content elements as well as content manipulation for TV enhancements,
- convergence of television, PC and the multimedia home network infrastructure and
- universal access to TV content anytime and anywhere as well as to personally created multimedia assets.

To present several scenarios would be beyond the scope of the book. Therefore only a few examples are explained and described in further detail. Table 9.1 gives an overview of how this service type relates to our key concepts.

9.1 Real-Time Content Manipulation

Many services can benefit from real-time content manipulation and depend upon which atomic unit manipulation techniques are used. On the multimedia asset layer this means the application of transcoding or adaptation techniques for enjoying digital TV on many different consumer devices. On higher layers it means customizing the content of multimedia assets (e.g. regionalizing advertisements).

Table 9.1. Relating intelligently presenting and interacting with content to our key concepts

Narrative Cube	
interaction	several interaction models possible
narrative	several narrative forms possible
asset	building of the synthetic space and its services (e.g. SVG models)
Digital Broadcast Item	
metadata definitions	application-specific metadata definitions (e.g. MPEG-7 for hyperlinked TV, SVG for small graphical add-ons), most higher-layer metadata building blocks are covered (e.g. narrative tools, service tools, object tools), asset tools as well as basic tools and vertical tools act as catalyst, emphasis on "after playout" metadata definitions, service compilation...
system architecture	consumer terminal focused implementations, involvement of multimedia home equipment (e.g. digital cameras, multimedia home servers), feedback channel actively utilized for many purposes (feedback services, home multimedia services), home multimedia gadgets...
dynamic behavior	focusing on end-consumer equipment, real-time feature extraction in production, real-time streaming of metadata, focus on interaction — storage — use —exchange of multimedia assets, instantiation of digital multimedia home item types and feedback channel item types...
local facilities	multi-platform development for home multimedia gadgets
communication model	manifold home networking protocols (e.g. Bluetooth), strong involvement of feedback channel protocols...
asset representation	service dependent, typical Internet representation models (e.g. image and file formats)...
Consumer Model Components	
strong involvement of consumer model components especially perceptive, interactive, intelligent, sociological elements	
Method and Technology	
wide range of consumer multimedia home technology, multimedia home service architecture, convergence of home platforms...	
Examples	
hyperlinked TV, real-time content manipulation, real-time content add-ons, intelligent customization of advertisements, distribution of multimedia assets to manifold consumer devices (e.g. PDA, PC), music application, content add-ons such as e.g. regionalized advertisement banners, broadcast show enhancing sound/audio effects, advanced content adaptation mechanisms, movie scene retrieval, narrative and semantic television intelligently personalized to consumer needs, active content elements as well as content manipulation for TV enhancements, convergence of television and the multimedia home network infrastructure, universal access to TV content anytime and anywhere as well as to self–created multimedia assets...	

Seeing it top-down, content manipulation means the design of multi channel applications independently distributed over any type of network: wireless, wired, broadband or narrowband. Emphasis is on "editing once — playout over any type of channel". The concept of MPEG-7 of universal multimedia access and the concept of MPEG-21 to provide item adaptation are entirely involved at this point. Again one layer higher, real-time content manipulation invokes narrative models with semantic descriptions to manipulate content in such form as the consumer would like to enjoy it. For live broadcasts real-time content manipulation means the introduction of real-time computing equipment doing the task of transcoding, metadata extraction and customizing content.

Example 9.1 (real-time content add-ons). Premarked places in video streams in the form of arbitrary objects can be utilized for displaying advertisement information. They can also be used for customizing video content in many forms. The user can exchange the actor's faces with his own or change the background scene to his own living environment. It is also possible to customize the content of complete virtual scenes.

Example 9.2 (intelligent customization of advertisements). There are many application scenarios for capturing metadata during production. Intelligent customization of advertisements addresses the issue of regionalizing, localizing and customizing static and camera-captured advertisements on banners. It is possible to involve the consumer context by regionalizing, personalizing or customizing advertisements by extracting information of advertisement positions from captured A/V materials and covering it by personalized, regionalized or customized new advertisement banners.

9.2 Convergence of the Multimedia Home

The term "convergence" is one of the major contemporary keywords. Convergence means merging different paradigms in the world of multimedia. The conceptualization and realization of platforms convey wireless protocol suites, home networking capabilities with emerging technology, unified data models and interactivity facilities. To satisfy consumers it is essential to build new hardware interfaces and extend functionality of simple remote controls towards more sophisticated and easier-to-use solutions. An interesting work dealing with the converging multimedia home has been written by P. Wallich [169].

The digital TV equipment can act as the first access point to multimedia home services and smart home solutions. Interoperability and integration are the hardest part. The efforts are supported by metadata and unified data models. They provide facilities to describe content and interchange it seamlessly over different distribution channels and multimedia home equipment. The underlying mechanisms are invisible to the end-user. Figure 9.1 shows a scenario where the digital TV equipment controls the fridge and room temperature, organizes multimedia assets and orders pizza.

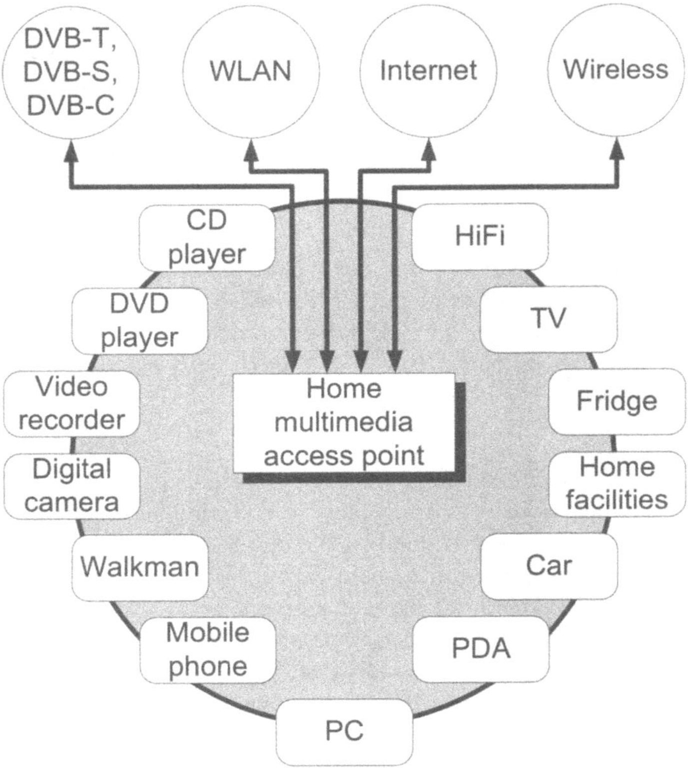

Fig. 9.1. The digital TV as first access point to multimedia home services (drawn very freely after [169])

Distribution of home multimedia services and the local TV equipment as a home access point for value-added services in a converged home network is an example for the converging home. The communication between devices such as PDAs, digital cameras, fridges and video/audio equipment enables interesting new services.

From the technology side, lightweight operating systems capable of running on small embedded and distributed devices enable home technology integration. Such an operating system empowers gadgets with the capability to run applications and communicate with each other over communication protocol stack implementations. A centralized home-server, storing holiday images, music files and videos supports the efforts of creating advanced service spaces in consumer homes.

9.3 Hyperlinked Television

Each movie can be hierarchically decomposed into its shots. Each shot has a representative frame or image, a so-called key-frame, with specific areas of interest. Specific areas of interest act as a hyperlink with an assigned action (see Fig. 9.2). The consumer clicks on a highlighted area of interest, and an Internet page is loaded or further descriptions e.g. about an actor appear on the TV screen. The idea of hyperlinked TV is not new and a quick search on the Internet results in various sources e.g. [101, 138, 28].

Fig. 9.2. Hyperlinked TV

To achieve this goal, each movie is annotated by its shot information, when it starts, stops and which key-frame is represented by the shot. Each key-frame contains regions with assigned actions, described by MPEG-7 descriptors. Both the actual movie content and its description are broadcast. While watching the movie, the consumer gets a signal if an annotated key-frame is present and based on his selections the key-frame and hyperlinks are displayed. Clicking or selecting hyperlinks triggers a certain action (e.g. opening an Internet shop to buy a gadget displayed in the movie). Figure 9.3 presents the overview of the implementation of the MPEG-7 metadata document for hyperlinked TV.

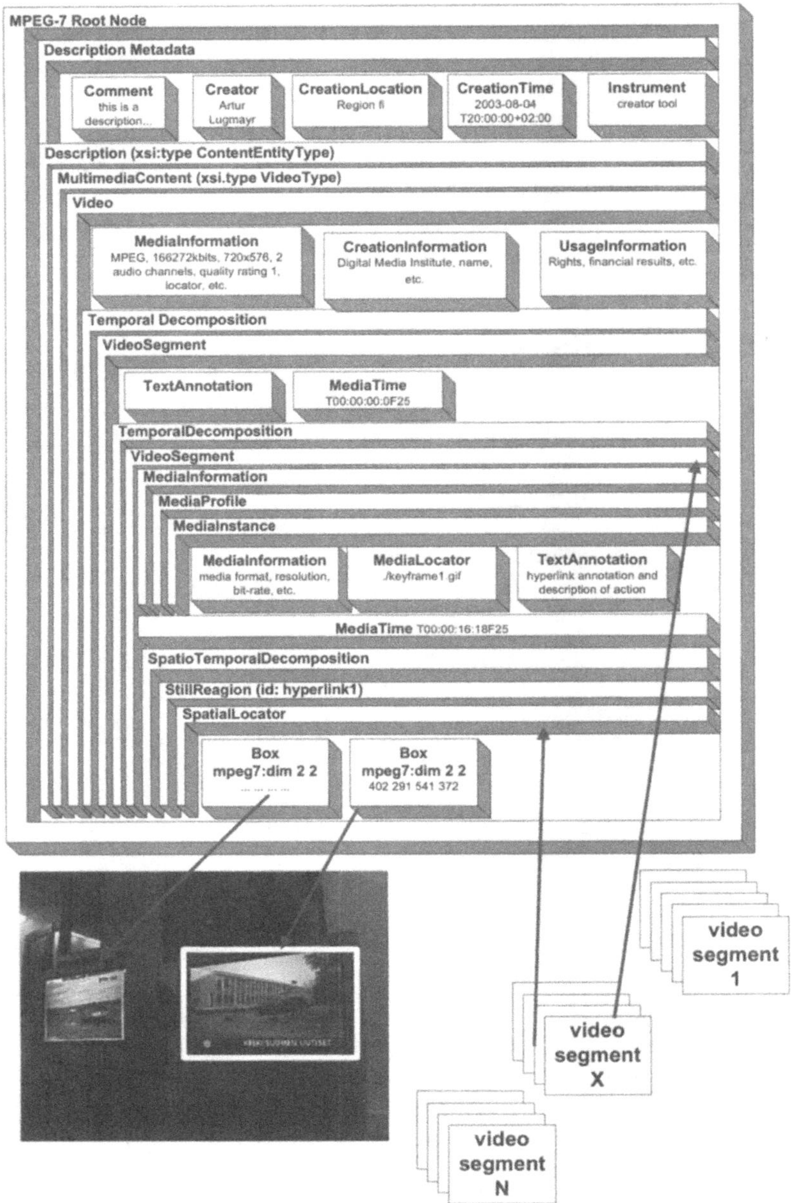

Fig. 9.3. Schematic description of the XML document for segmented TV as shown in Fig. 9.2

10

Consumer Profiling and Personalization

The Merriam Webster Online Dictionary has the following definitions for personalization: "[...] to make personal or individual; specifically: to mark as the property of a particular person[...]" [131].

Personalization is one of the defining concepts of a contemporary multimedia platform. As evident from the dictionary definition personalization conveys the meaning of making something personal or individual. Within the context of multimedia, personalization involves both multimedia content (video, audio, hypertext, images, etc.) and presentation and interaction context (user interface, etc.) personalization. Table 10.1 gives an overview of how this service type relates to our key concepts.

In networked multimedia content and services are provided in a distributed media environment with various content access and service types. As we know, broadcast multimedia belong to networked multimedia. Contemporary broadcast multimedia environments involve services, which can include both broadcast and interactive multimedia content and presentation and interaction contexts. The contexts are typically in the form of user interfaces to multimedia content or applications providing interactivity. The services can have either local or nonlocal interaction capabilities. Through digitalization, television is becoming an avenue for the realization of personalized broadcast multimedia services [142].

With the digitalization of television a wider variety of multimedia content types together with services is becoming available for the consumer. This information overflow underscores the need for new methodologies for the consumer to efficiently seek content and services in his or her own interest. This chapter seeks to elucidate how advanced metadata-driven personalization can be realized in digital TV.

Metadata is a key concept in broadcast multimedia personalization as it provides a descriptive framework for the complex data to be personalized. This enables personalization related methodologies such as data matching and filtering based e.g. on user preferences.

Table 10.1. Relating consumer profiling and personalization with content to our key concepts

Narrative Cube	
interaction	services with varying interaction models may be personalized
narrative	services with varying narrative forms may be personalized
asset	arbitrary broadcast multimedia assets
Digital Broadcast Item	
metadata definitions	Consumer action history data (MPEG-7 / TV-Anytime), consumer profiles (MPEG-7 / TV-Anytime), multimedia asset descriptions (MPEG-7 / TV-Anytime / DVB-SI / W3C metadata)...
system architecture	focus on the delivery of up-to-date metadata assets to the consumer terminal, provision of intelligent processing facilities for different metadata assets (e.g. consumer profiles and content descriptions) and on the efficient sharing and communication of personalization metadata among different actors of the personalization service schemes. ...
dynamic behavior	Real-time provision of broadcast channel metadata assets, embedment of descriptive metadata into both broadcast and feedback channel content assets, instantiation of different DBI types within personalization service schemes, real-time processing of metadata assets, real-time capture of consumer TV viewing behavior and their distillation into behavior pattern metadata...
local facilities	a consumer home network and e.g. home server connected to the consumer STB serve as depository of permanently stored consumer metadata assets...
communication model	sharing of personalization metadata assets through DBI exchanges over digital TV broadcast and feedback channels...
asset representation	all asset types present at the consumer-end of a digital TV asset-cycle, MPEG-2 TS A/V content and W3C XML-based hypermedia content (e.g. HTML)...
Consumer Model Components	
consumer is at the center of the personalization service schemes, use of consumer behavior pattern metadata describing asset consumption patterns...	
Method and Technology	
embedment, creation, storing and processing of metadata assets related to personalization service schemes throughout the digital TV asset lifecycle by using the DBI methodology, a mechanism for realizing seamless personalization service schemes in digital TV...	
Examples	
personalization of broadcast multimedia assets of arbitrary types, TV program recommendation generation service, customization of value-added digital TV service user interfaces and content...	

Personalization in digital TV has three application areas: personalization of the standard A/V broadcast content in the form of TV program recommendation generation functionality, value-added content personalization and value-added content presentation context and interaction context personalization. The latter two are realized as content and user interface personalization in value-added applications. Personalization of standard A/V broadcasts enables consumers to conveniently seek TV programs that meet their individual preferences. Value-added service personalization helps consumers to experience value-added services with user interfaces and interactive multimedia content adapted to their personal needs and requirements [142].

Table 10.2 gives examples of digital TV personalization services in the three application areas.

Table 10.2. Examples of digital TV personalization services

Target of Personalization	Example Services
broadcast A/V content	*searching for and filtering available TV program content* Matching consumer profile and available content metadata *profiled compilation and recording of TV Broadcasts* e.g. compilation of personalized news broadcasts to the consumer STB
value-added content	*content adaptation for consumer groups or individuals* adaptable content object (text, images) sizes for people with poor eyesight individualized content-based on consumer or group profiles
value-added contexts	*user interface adaptation for consumer groups or individuals* adaptable user interface complexity for impaired user groups individualized user interfaces on consumer or group profiles

10.1 A Metadata-Driven Approach to Digital TV Personalization

The use of a metadata-driven and digital broadcast item-based methodology provides a unified approach to the conveyance, processing and storage of multimedia assets within the personalized digital TV value-chain.

Several digital TV and Internet metadata standards include support for personalization in the form of data structures specifically defined for the purpose and otherwise applicable in the personalization domain.

TV-Anytime metadata [87, 88] includes a number of descriptor schemes inherited from MPEG-7 related to personalization. The third category of TV-Anytime metadata, consumer metadata, includes usage history data, annotation metadata and user preferences. Usage history metadata is implicitly generated when the consumer views and browses content. Examples of explicitly generated annotation metadata are bookmarks made by the consumer. The usage history DS (as defined by MPEG-7) is used for gathering usage information over periods of time. The collected usage history provides a list of the actions carried out by the user during an arbitrary observation period. This action database can be subsequently used, e.g., by automatic analysis techniques for generating user preference information.

User Preferences enable the description of a user's preferences related to the consumption of multimedia assets. User preference descriptions can be correlated with multimedia content descriptions to search and filter content desired by the consumer. Matching of user preference and content descriptions facilitates accurate personalization of multimedia content and services. In addition to the specifically personalization related consumer metadata, TV-Anytime content and instance metadata can be utilized in digital TV personalization. From DVB-SI the standard TV broadcast description metadata, e.g. from the SDT table, can be utilized in personalization services.

W3C metadata standards (e.g. *resource description framework (RDF)*) originally developed for the Web-based hypermedia content description are applicable for value-added digital TV services.

Figure 10.1 shows an abstract description of metadata-driven, digital broadcast item-based personalization in digital television. The consumer and the digital TV STB are in the center of the figure. Digital multimedia assets (content, services and metadata annotating them) are provided through the broadcast and feedback channels of a digital TV system. Digital broadcast items serve as logical containers for personalization related descriptors and content. The user STB is equipped with a digital broadcast item browser to provide personalized access to the multimedia assets. Consumer metadata is logically contained in digital broadcast consumer items available through the consumer home multimedia network e.g. from a home server. All data within this approach are logically exchanged through digital broadcast item exchanges.

The DBI browser provides the consumer with personalized broadcast A/V content, value-added content and value-added content presentation and interaction contexts. This is done through consumer profile and content metadata matching. Consumer profiling can be based on either explicit collected or implicitly generated consumer preference metadata. The generation of consumer preference metadata from implicitly collected action history metadata requires the use of advanced metadata processing methodologies.

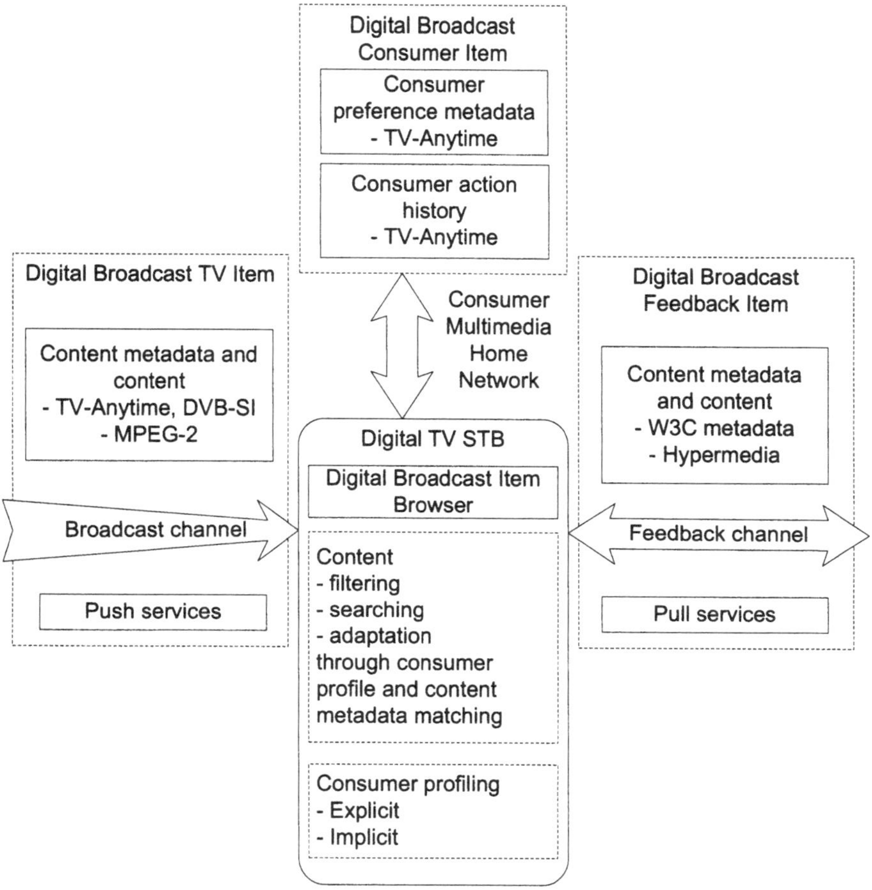

Fig. 10.1. A metadata-driven approach to digital TV personalization

10.2 Advanced Processing of Personalization Metadata

Knowledge discovery in databases (KDD) is an information technology tool for refining useful knowledge or patterns from the rapidly growing volumes of ubiquitous data. KDD is a young discipline with many potential application areas. In some application domains KDD has already been proven to be an important tool. One of these domains is multimedia personalization where it is the key tool in consumer profiling for metadata-driven approaches to digital TV personalization.

The KDD process has been given the following definition [70]: "The non-trivial process of identifying valid, novel, potentially useful and ultimately understandable patterns in data."

U. Fayyad et al. have described KDD through a process-centric framework consisting of nine steps from data selection to pattern interpretation or evaluation [69, 70]:

1. *familiarizing the application domain:* learning the concepts and properties of the application domain including the goals of the application;
2. *target dataset selection:* selection of a dataset or focusing on a subset of variables or data samples on which knowledge discovery is to be performed;
3. *data cleaning and/or preprocessing:* removal of noise and outliers, deciding on handling missing data fields and deciding on database system issues such as data types, schemata and mapping of missing or unknown values;
4. *data reduction and/or projection:* finding useful features for representing the whole data set and, depending on the task at hand, using dimensionality reduction or transformation methods to reduce the effective number of variables under consideration or find invariant representations for the data in comparison to the original data set;
5. *choosing the function of data mining:* decision for the purpose of the model derived by the data mining algorithm (e.g. summarization or classification);
6. *selecting the data mining algorithm(s):* selection of method(s) to be used in searching for patterns in the data, e.g. deciding which models and parameters may be appropriate and matching a particular data mining algorithm with the overall criteria of the KDD process (e.g. we may be more interested in understanding the model than its predictive capabilities);
7. *data mining:* searching for relevant patterns in a particular representational form including classification rules, regression and clustering;
8. *interpretation:* interpretation of the discovered patterns and possibly reverting back to any of the previous steps, visualization of the patterns, removal of redundant or irrelevant patterns and translating the relevant ones into human-readable form depending on the application;
9. *using the discovered knowledge, the patterns:* incorporating the patterns into an external process where the knowledge is to be utilized, taking human-initiated actions based on the knowledge, or simply documenting and reporting the knowledge to relevant parties.

In Fig. 10.2 an abstract description of the KDD process is given as a flow diagram.

It should be noted that steps 5 to 7 described earlier are incorporated into the *data mining* step of the diagram and that the process has the option of reverting to any previous step after one step is finished. This makes it possible to iteratively search for optimal pattern acquisition strategies within the process flow.

After defining KDD let us return to considering its application in digital TV personalization. The domain within personalization relevant to KDD is consumer profiling. Typically the goal is to seek patterns of interest in usage

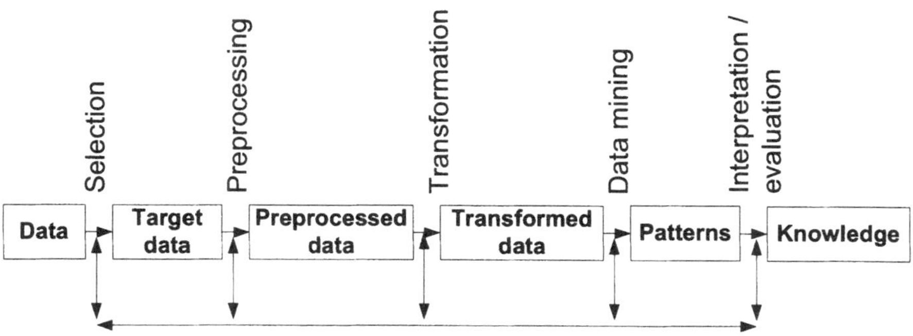

Fig. 10.2. The overall KDD process as adapted from Fayyad et al. [69]

history data collected from consumers. The patterns define consumer preferences and interests in different content categories in specific usage contexts. To attain this goal, the described nine-step KDD process can be applied.

When looking into the KDD process in consumer profiling, the target data set is standardized (e.g. TV-Anytime) TV usage history metadata collected either implicitly or explicitly from consumers. The target data set should not require any preprocessing, as it is automatically generated within well-defined and validating (e.g. XML-based) metadata frameworks. However, the dataset could still be processed to detect and remove any missing data from the personalization process. In the reduction step, some numerical attributes of the user history metadata may be transformed into ranges and intervals.

Many data mining algorithms are suitable for consumer profile metadata generation. As an example, rough sets theory is ideally suited for data mining applications with vagueness and uncertainty in the initial data. This is typically the case with data describing human behavior. The theory of rough sets and its application in knowledge discovery are well described in [145] and [175], respectively.

"Practically all rough sets apps can be called knowledge discovery apps" — W. Ziarko in [175]. As pointed out by Ziarko, the use of rough sets theory in information sciences is typically related to a knowledge discovery related application or application area. Basically, the applications of rough sets theory in data mining involve the collection of empirical data with the aim of building classification models from the data.

The process corresponding to this general application type has been characterized as concerning the acquisition of decision tables from data. This is followed by the analysis and simplification of the decision tables by the identification of attribute dependencies, minimal non redundant subsets of attributes and finally minimized rules or patterns in the data mining sense of the word.

Personalization patterns or rules generated with rough sets data mining from previous user behavior have by default different levels of validity (depending on the degree of persistence of a particular pattern corresponding to a rule) corresponding to different levels of recommendation for the user in the case of a rule match. The possible vagueness or uncertainty in the original human behavior data (e.g. resulting from contradictory behavior) is thus handled automatically through the different levels of validity of the rules, a significant benefit compared to some other data mining methodologies. The discovered consumer profile knowledge can be applied in consumer profile-based personalization. A possible use-scenario is the matching of available content metadata against consumer preferences to generate online recommendations for available TV programs or to provide automatic program recording or compilation functionality.

10.3 Case Scenario: TV Program Recommendation Service

Another scenario for metadata-driven personalization in digital TV is a service providing recommendations for available TV programs for the consumer. Fig. 10.3 gives an abstract description of the service. Basically, it enables distilling a TV viewer's past behavior patterns from the TV usage history data to help recommend viewing choices for currently available TV A/V broadcasts through a metadata comparison. TV usage history metadata is collected implicitly and a suitable metadata processing methodology (e.g. rough sets-based KDD) is employed to generate consumer viewing patterns from the raw usage history. In the content recommendation service, the generated preferences are matched against the currently available broadcast TV programs and recommendations are visualized on the consumer TV screen for each available service.

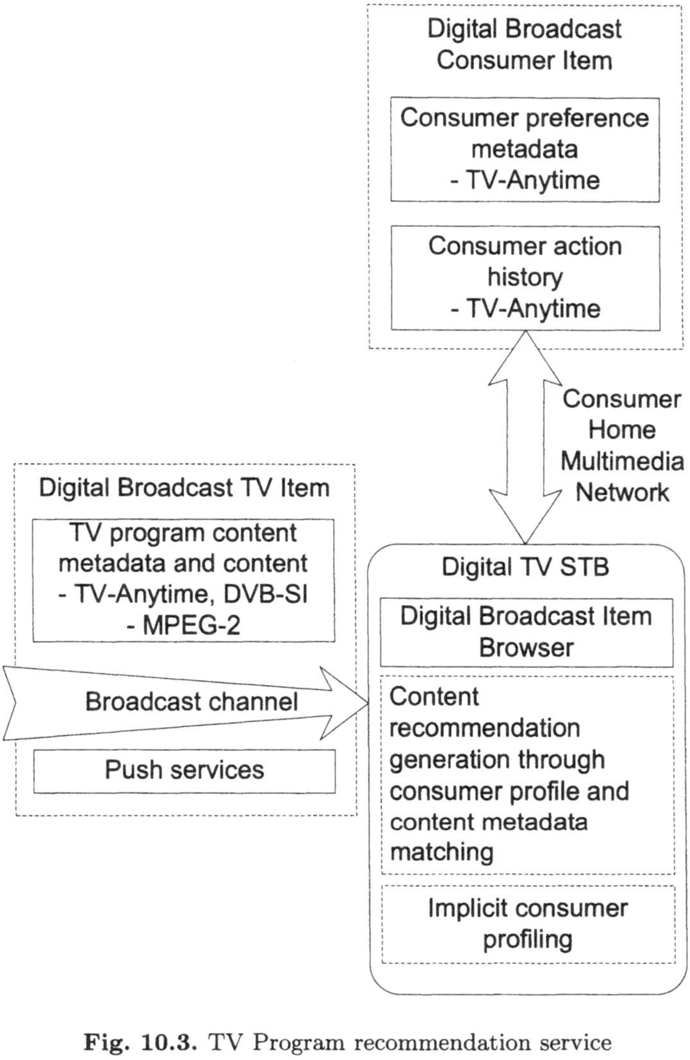

Fig. 10.3. TV Program recommendation service

11

Ambient TV

TV anytime, anywhere, anyhow and according to the consumer's means! This does not only relate to TV in its current form for distributing monolithic assets to consumers. Seeing TV and all its newly emerging forms converging towards a service space shifts paradigms tremendously. TV is a medium for distributing any arbitrary services. The European Union developed therefore the concept of ambient multimedia [95] as goal for their future project initiatives.

This means first embedding the TV into multimedia service spaces and vice versa surrounding the TV with new user-centered and converged service spaces. Therefore TV is not only a simple medium for distributing monolithic assets — it is a multimodal, interactive, mobile and cooperative service space. Table 11.1 gives an overview of how this service type relates to our key concepts.

11.1 Creation of Ambient Multimedia as Concept for Digital TV

The concept of ambient services is illustrated in Fig. 11.1. On a higher abstraction level communities (digital or real ones) act in the real world in a certain context. The real world is the environment where humans are actually using the system. This might be at home, in the car or on the mobile on-the-move. The context is more precise and describes surrounding conditions, such as user mood, available interaction facilities, situational information and environmental parameters (e.g. weather conditions, light conditions, noise level, degree of possible interaction modes).

To realize such service types, four components are essential:

- *ambient intelligence* is the catalysis of intelligent distributed systems, context awareness, intelligent interaction, intelligent interoperability and location awareness towards a service space by actively involving digital cultures and human interaction;

Table 11.1. Relating ambient TV to our key concepts

Narrative Cube	
interaction	strong focus on strong and hybrid interactivity
narrative	no direct impact on the narrative
asset	support for any type of assets aimed at consumer homes
Digital Broadcast Item	
metadata definitions	consumer terminal metadata, metadata definitions for communication between embedded devices, digital items for the exchange within homes...
system architecture	embedded devices, disappearing hard- and software, systems that recognize their environment, intelligent equipment...
dynamic behavior	highly responsive to consumer inputs, relates consumer and systems to one entity...
local facilities	strong focus on embedded devices...
communication model	in-home communication, strong involvement of wireless protocol types...
asset representation	assets are partly hidden from the user, assets manifest in services surrounding the consumer, multimedia home equipment...
Consumer Model Components	
very strong focus on several aspects of the consumer model	
Method and Technology	
pervasive computation, ubiquitous computation, intelligent algorithms, software-based artificial intelligence, ambient intelligence, service oriented data models, sensor-based environments...	
Examples	
automatic response to consumer surrounding environments, digital TV as access point to services, vision-based input devices, sensor-based environments...	

- *ambient culture* for creating an intelligent, context-aware service space for social networks or communities by building an immersive digital environment for aesthetic, domain-distinctive and interactive multimedia assets capable of actively involving user responses;
- *disappearing hardware and software* refers to a pervasive and ubiquitous networked system design by putting sensors and devices everywhere and making them invisible to the consumer;
- *service-oriented data models* as an abstraction of content and metadata in the form of multimedia assets. Several entities are strictly separated from their applicable service types by applying fundamental metadata concepts.

Fig. 11.2 shows a scenario for an environment where the user is surrounded by sensors and actuators. Sensors obtain environmental data and actuators perform actions according to the input. It is possible to watch TV location independently. Sensors locate persons and person groups and these data are

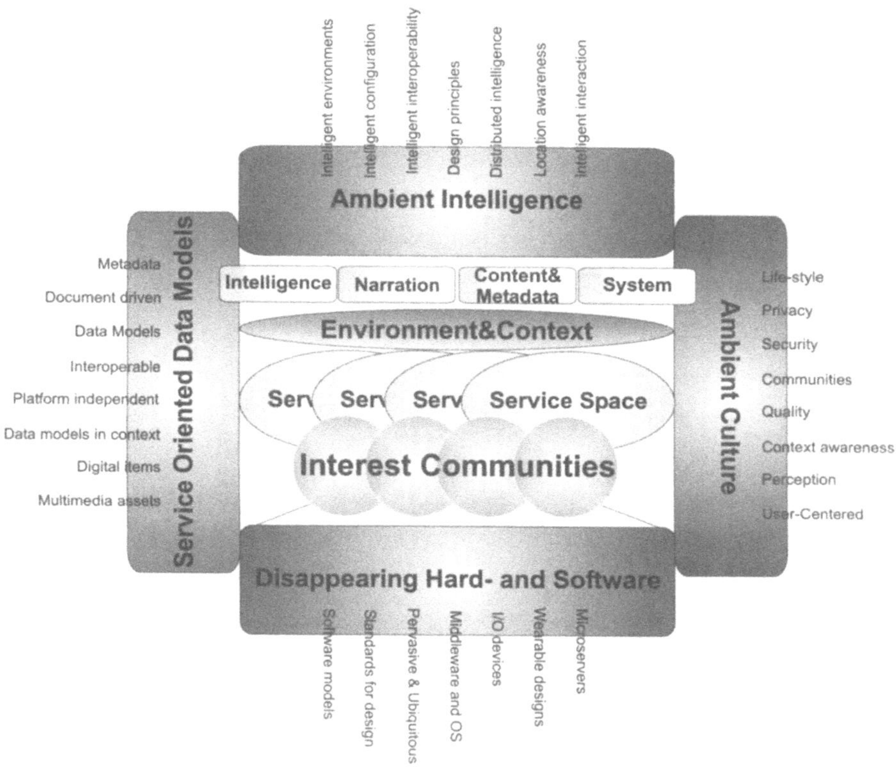

Fig. 11.1. Ambient TV as concept

utilized to adjust sound rendering within the rooms as well as putting the current broadcast TV show on the screens nearest to the user.

Considering the digital TV set as an access point for services, it acts as a service portal for the multimedia home: with the help of the remote control it is possible to control the fridge, adjust living-room lighting, order shopping goods, collect data from gadgets (e.g. digital photo camera) and gain control over multimedia home equipment and many other digital home devices. Due to the rapid changes in technology development, it is hard to predict and anticipate the process of determining which one might find the way into consumer's homes. Digital TV is one of the factors that might lead to breakthroughs in the development of home technology and multimedia home integration. Wireless design and lightweight operating systems are predominant.

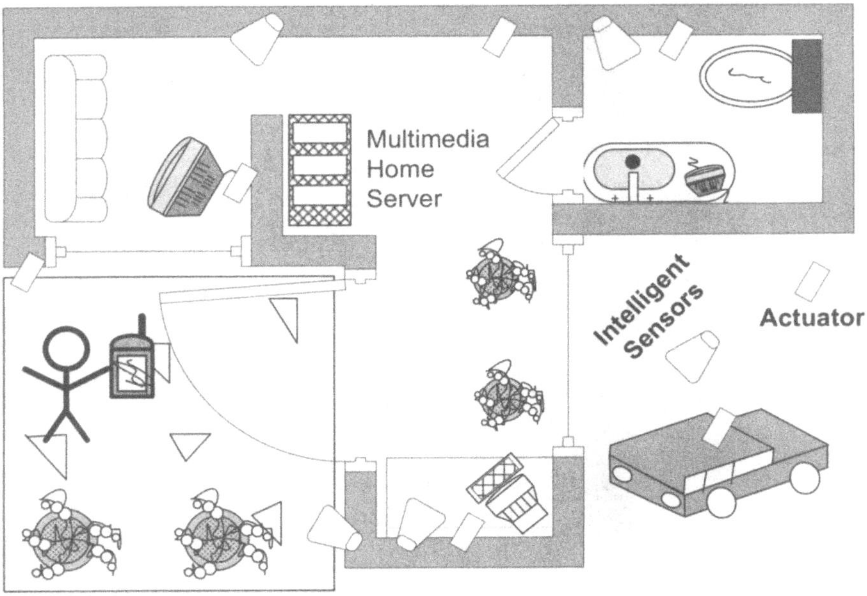

Fig. 11.2. TV everywhere at home

12

Other Application and Service Scenarios

There are many other interesting and challenging applications on the newly emerging platform of digital TV. To enumerate all of them as well as to envision where digital TV might go in the future is rather complex. Still, in the following there are some service scenarios and applications that might emerge on this very powerful platform. With these additional scenarios we are able to point out some interesting application areas of digital TV that are not covered in previous sections of the book.

12.1 Evolving Wireless Story — Big-Brother TV

Watching others, private lives and being seen on TV have already proved "Big-Brother TV"-like shows to be a success story in analog TV. Digitalization of television provides new opportunities for creating shows similar to Big-brother TV. Camera equipment gets smaller and smaller and people are able to create their own assets for presenting themselves and exhibiting their private lives to the public.

12.2 Digital Soap Operas

The concept of narrative TV has been exhaustively presented within the scope of this book. But soap operas are central components of current TV program schedules. Soap operas have to be produced fast, as cheap as possible and have to fit to the consumer's desires or life. Digitalization might enable many different scenarios: digital production helps to reduce costs and produce soap operas very rapidly; soap operas can be personalized via personal profiles from the consumer; and the consumer is enabled to create his own digital soap opera as an evolving story composed of 3D graphics.

12.3 Live Sports Event Tracking

Digitalization enables spicing up sports events on digital TV with informational services, event analysis, multiple camera perspectives, betting, add-on services and game content. Screen-overlaying panels show additional information in textual or graphical form for live sports events. Currently only the broadcaster has the technology to alter A/V content. In future the composition of the live sports event might be selected by the consumer. The consumer is enabled to select information types he desires, including graphical information and compile the scene as he wishes.

12.4 Live Tracking

Most recent news broadcasts applied the concept of embedded reporters. Reporters are equipped with real-time broadcasting equipment and create their content from a site live. The concept itself is interesting, as the consumer is virtually embedded into the live event and sees it from a centered perspective. Future technology might enable him to explore the live scene and discover the happenings by himself.

12.5 Wearable TV

Merging TV with real-world components and embedding and surrounding the consumer in a complete virtual television environment might be envisioned by this scenario. Mobile devices and wireless protocols enable the consumer to move freely wherever he wants. Increased processor performance and transmission bandwidth support wearable television.

12.6 3D-TV

Very many experiments around 3D-TV have been already made. 3D supports not only the production processes, but is also a feature for consumer terminals. Advanced display types immersing the consumer enable fascinating new scenarios for digital TV. Computer game-like television shows and virtual reality-like interaction models make digital TV an unforgettable experience.

12.7 Display Technology

The simplest and most commonly wide-spread argument for digital television is its enhanced A/V capability. The consumer devices get more powerful and digital TV offers high-quality television streams. This also requires advanced display technology at home to present multimedia assets in their appropriate form. Higher pixel resolutions, increased color depth and larger display types round the television experience up. One example for a novel display technology that might come to consumer homes some day is the fog-screen developed by Ismo Rakkolainen (see Fig. 12.1).[1]

Fig. 12.1. Novel display type: a fog screen — alias walkthrough or airborne display — displays images that appear to float in a layer of thin air (© Fog Screen Inc., www.fogscreen.com)

12.8 Vision-Based Digital TV Mouse

Why not merge paradigms? The mouse as an input device for the PC is indispensable. A Web-camera as an input device plugged into a digital TV consumer device enables the technology to obtain parameters from a marker in the consumer's hand. These positions are transmitted to an application that controls the movement of a virtual cursor on the digital TV screen. Mouse

[1] We would like to thank Ismo Rakkolainen from Fog Screen Inc., Tampere, Finland (`http://www.fogscreen.com`) for the provision of his materials and the invention of this excellent, innovative display type.

clicks are enabled via moving the marker held by the consumer more rapidly or forward-back.

A vision-based digital TV mouse also has its application area for persons with special needs. Figure 12.2 shows the principle of a vision-based digital TV mouse.[2]

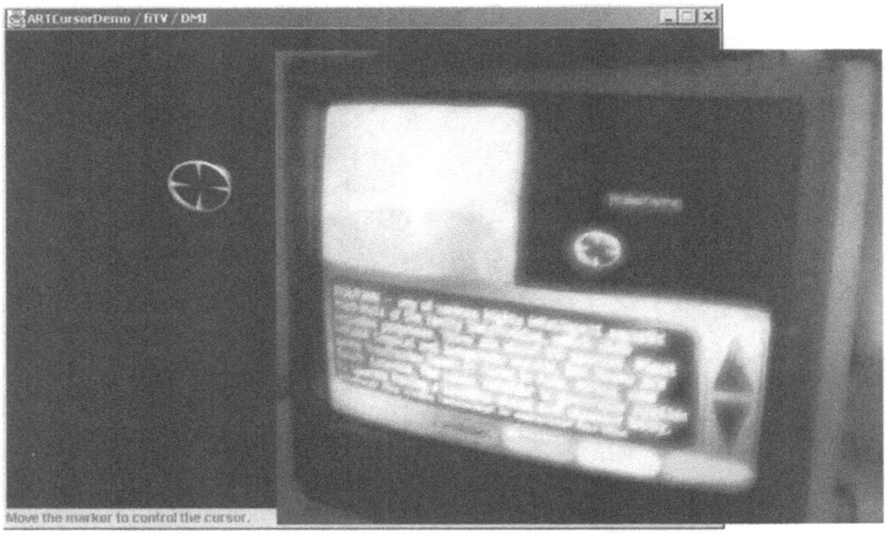

Fig. 12.2. Vision-based digital TV mouse (PC version and digital TV version)

12.9 Digital TV as Personal Life Organizer

Simple applications currently performed on the PC might be interesting to migrate to the platform digital TV. They are not in direct correlation with the actual purpose of watching television. It is a fact that the TV set has a central role in consumer homes. Digital TV equipment is also a small-scale PC in the consumer's home. And this small-scale PC is capable of taking over tasks from the PC. There is only one big advantage: the maintenance efforts for the PC are rather high, whereas a STB is an embedded device. This makes it an excellent platform for the implementation of PC-based applications, such as personal life organizers, address books, andmultimedia home databases among others.

[2] The implementation of the vision-based digital TV mouse has been made by Jussi Lyytinen at the Digital Media Institute, Tampere University of Technolgoy, Tampere, Finland within the future interaction TV project (`http://www.futureinteraction.tv`).

13

Road Ahead in Broadcast Multimedia

To better understand the tremendous changes in broadcast multimedia that are lying ahead, we look a bit into the history of broadcast television. These observations enable us to conceptualize the overall trends in television and to map the future of broadcast multimedia.

There are four key factors in broadcast multimedia that describe its evolution in the historical context:

- *revolutions* represent the overall evolution of broadcast multimedia from initial innovation to a futuristic TV. Looking into the past and envisioning the future of broadcast multimedia, the following revolutions are present: *initial invention of TV*, *TV as a mass product*, *diversification of broadcast TV content*, *emergence of interactivity in TV*, *media convergence* and *TV as an immersive experience*;
- *aspects* define the basic paradigm changes that lead to revolutions in broadcast multimedia. One specific example is the basic technology evolution from analog to digital TV. Aspects define paradigm evolution on an abstract level. Three general aspects, *technology*, *service space* (analogue, digital and biological) and *perception* are considered here;
- *steps* are small innovations changing and enhancing the television experience in the domain of the three general aspects (e.g. adoption of the video cassette recorder in analog television);
- point of *time* defines changes coming from outside broadcast multimedia (e.g. societal changes).

Considering the three aspects (service space, technology and perception) we can ask what they have meant and what they will mean in the context of broadcast multimedia. It is clear that the three aspects are closely connected to the "revolutions", "steps" and "time" factors in the evolution of broadcast multimedia. Figure 13.1 gives an overview of how the aspects and other factors are related to each other.

Technology. "TV" emerged as a technological invention in the early twentieth century (initial invention of TV) and found its way into the consumer

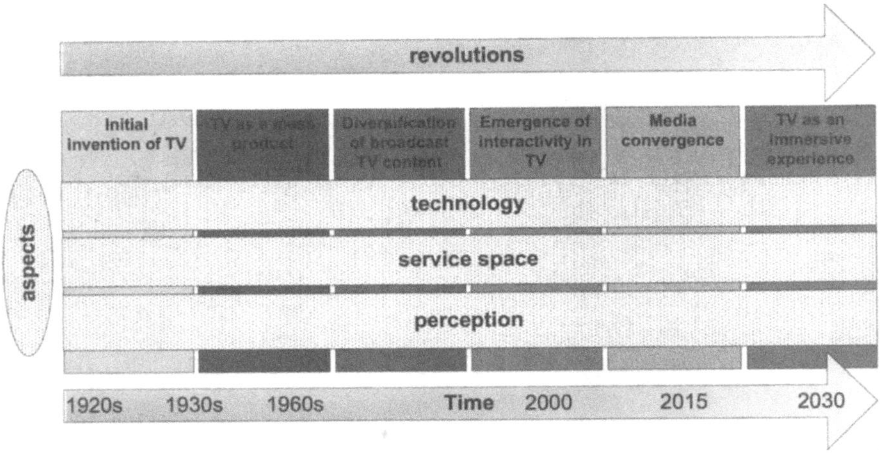

Fig. 13.1. Factors in broadcast multimedia

homes of the industrialized world beginning in the 1930s. Still, the adoption of TV as a mass-product available to all consumers required three more decades. Along with commercialization of color TV in the 1960s television became a settled mass-product and a wide-spread technology. The time span between invention and wide-spread technology adoption required approximately 40 years (TV as a mass-product). From the year 1960 to the year 2000 there were only a few changes in TV technology visible to the consumer. TV technology was gradually enhanced and additional features such as teletext, consumer recording devices, more sophisticated production systems and payment models were introduced. In conclusion, there was little visible change in TV technology for over 30 years.

Broadcast television built on the premise of an analog broadcast signal began to face severe limitations in the late twentieth century as more and more TV content became available. On the other hand consumers began to expect more from electronic media after introduction of digital mobile communications and the Internet to a wide audience in the 1990s. While the technology of analog TV matured, common content and interactivity modalities changed rapidly. Consumers got used to the instant interactivity of the Internet and to the mobility of GSM phones. As an answer to this call, digitalization of television is finally under adoption in the early twenty-first century. In addition to the digitalization of the broadcast signal, digital TV provides wholly new service modalities for TV in the form of value-added services and advanced interactivity (emergence of interactivity in TV).

Currently another technological trend is emerging. Digital information will be available anytime, anywhere and anyhow for these consumers entitled to access it. Only the convergence of platforms, distribution channels and service spaces will satisfy this trend (media convergence). As an ongoing trend it will be one of the main technological challenges for the coming years. Multiple platforms, channels and content models require unified access schemes and content protection techniques. There are two trends emerging: the boundaries between public and private information systems will break down and increased use of digital rights management and content protection mechanisms. The probable convergence path for digital TV technology is the move towards increased mobility and the adoption of Internet-technology within the broadcasting value-chain.

Service Space. Looking at the service space (or to value-added delivered through broadcast multimedia to consumers) we see tremendous historical and future changes. Looking first at the content side or content revolution, we state that the service space did not dramatically change until the 1960s. Public or centralized broadcasters created broadcast shows, mostly aligned with general societal and political considerations. The 1960s meant the breakpoint for the evolution of broadcast content due to the general liberalization of society and culture and to the emergence of the consumer society. This decade is a magic point, where TV settled its position as a widely accepted mass-product, while the content that was delivered to TV began to dramatically change (diversification of broadcast TV content).

As time passed the television content and services diversified. The basic technology visible to the consumer remained analog and could no longer face the challenges brought up by the accelerating content revolution. This development coincided with the emergence of other digital media. Consumers got used to the capabilities of the Internet and digital mobile media. It is obvious that at the beginning of the twenty-first century TV had to change or its position as the leading platform in electronic mass media would be taken by some other medium. The digital revolution in TV started by replacing the basic technology of delivery and reception with digital systems.

As noted earlier, the megatrend in information technology is related to the convergence of platforms, services and content. The emergence of broadcast multimedia is the precursor of the creation of a unified service space. In more practical terms this development is visible in the emergence of digital TV. As service types typically associated with the traditional Internet became possible in digital TV due to digitalization and the availability of an interaction medium a need arose for unified approaches to manage system and content complexity as well as the diverse value-added services (emergence of interactivity in TV). The approach of this book represents a novel solution to this challenge in the form of a unified metadata-driven design approach for broadcast multimedia.

Convergence implies the utilization of metadata throughout the digital path in the value-chain of broadcast multimedia. Metadata is the answer to content revolutions and changes demanded by the consumer. It is the key growth-inducing factor for the expanding and diversifying service space of digital TV. Services and TV once again emerge as the leading mass-media for consumers. Digital TV would not be the major platform rather than a substation medium for others.

Perception. In the early days of analog TV, the theater and books formed other synthetic environments available to consumers. It is important to understand the paradigm change that was ongoing at that time. To increase perception consumers should get immersed into the medium. "To immerse" means "to plunge into something that surrounds or covers [... or...] to plunge or dip into a fluid" [131]. Especially in computer science it is used for describing how a consumer gets surrounded by virtual scenes. We use this term in the context of digital TV for surrounding and embedding the consumer in a space of services. The ways and media for presenting or getting immersed in leisure content were changing to a different new one. Radio created a virtual aural environment, similar to analog TV, which created a virtual visual environment. TV was a novel medium for obtaining content as it is in many ways now. TV does not completely substitute another medium — it will stay as *one* among the others.

Convergence means interconnecting multiple media in order to create a service space in which the consumer is immersed. The key question is — if — in future television or its derivative will be the standard medium for immersing consumers in content (TV as an immersive experience) or will it be some other medium? In the latter case, TV would play the role of the theater or books in the early 1920s. Some new medium would come up for providing this type of service space to consumers. Perhaps this medium will be a biologically inspired environment as currently presented in various Hollywood movies. Humans are interconnected via strings of bio-cables stimulating nerves to directly present content to the human perceptive system. We do not yet know which basic technology will emerge as the one to make the borderline between real-world and fictive world disappear for common consumers.

In our forecast, the first mega-trend in the twenty-first century is digital convergence — metadata being a key factor in it — whose adoption in digital TV establishes television as the precursor of a unified media environment. If television gains a dominant role in the attained unified service space, the next big step, immersiveness, will also most likely be built using television as a basic foundation. However, it is important to state that this step will make the medium currently known as TV obsolete. New technology based on biological models and direct stimulation of the senses will take its place.

14

Abbreviations and Acronyms

A/V	Audio/Video
AAF	Advanced Authoring Format
AIT	Application Information Table
API	Application Programming Interface
ATSC	Advanced Television Systems Committee
AU	Access Unit
B2	Business-to-Consumer
B2B	Business-to-Business
B2C	Business-to-Consumer
BIM	Binary Format for MPEG-7 Description Schemes
BSP	Broadcast Service Provider
CA	Conditional Access
CAT	Conditional Access Table
CATV	Cable TV
CCIR	MPEG-2 studio quality
CMHN	Consumer Multimedia Home Network
CPU	Central Processing Unit
CRID	Content Reference ID
D	Descriptors
DBCIT	Digital Broadcast Creation Item Type
DBCMIT	Digital Broadcast Consumer Multimedia Home Item Type
DBCMO	Digital Broadcast Multimedia Home Object
DBCO	Digital Broadcast Creation Object
DBFIT	Digital Broadcast Feedback Item Type
DBFO	Digital Broadcast Feedback Object
DBGIT	Digital Broadcast Generic Item Type
DBI	Digital Broadcast Item
DBIM	Digital Broadcast Item Model
DBO	Digital Broadcast Object
DBSIO	Digital Broadcast Service Item Object
DBSIT	Digital Broadcast Service Item Type

DBTVIT	Digital Broadcast TV Item Type
DBTVO	Digital Broadcast TV Object
DDL	Descriptor Definition Language
DF	Directory Facilitator
DI	Digital Item
DIA	Digital Item Adaptation
DRM	Digital Rights Management
DRMS	Digital Rights Management System
DS	MPEG-7 Descriptor Scheme
DSM	Digital Storage Media
DSM-CC	Digital Storage Media - Command and Control
DVB	Digital Video Broadcasting
DVB-C	DVB Cable
DVB-HTML	DVB-HTML application
DVB-J	DVB-Java application
DVB-S	DVB-Satellite
DVB-SI	DVB-Service Information
DVB-T	DVB-Terrestrial
EBU	European Broadcasting Union
EE	Execution Engine
EPG	Electronic Program Guide
ES	Elementary Stream
FP	File Package
FUU	Fragment Update Units
GEM	Globally Executable MHP (GEM)
GXF	General Exchange Format
HAVi	Home Audio/Video Interoperability
HDTV	High Definition Television
HTML	HyperText Markup Language
HTTP	HyperText Transfer Protocol
IEC	International Electrotechnical Commission
IP	Internet Protocol
IPR	Intellectual Property Rights
ISDB	Integrated Services Digital Broadcasting
ISDN	Integrated Services Digital Network
ISO	International Standardization Organization
ISP	Interactive Service Provider
JavaTV	Extension of the Java platform for digiTV
JDOM	Java-based Document Object Model for XML documents
KDD	Knowledge Discovery in Databases
KLV	Key-Length-Value
MAA	Multimedia Asset Adaptation
MDL	Multimedia Description Language
MDS	Multimedia Description Schemes
MHP	Multimedia Home Platform

MIME	Multipurpose Internet Mail Extensions
MMS	Multimedia Message
MP	Material Packages
MP@ML	Main Profile at Main Level
MPEG	Moving Picture Experts Group
MXF	Material eXchange Format
NIT	Network Information Table
NVOD	Near Video On Demand
OCAP	OpenCable Applications Platform
PAL	Phase Alternating Line
PAT	Program Association Table
PDA	Personal Digital Assistant
PE	Presentation Engine
PES	Program Elementary Stream
PID	Packet Identifier
PM	Protection Management
PMM	Protection Management Mechanisms
PMT	Program Map Table
POP	Post Office Protocol
PPV	Pay-Per-View
Pro-MPEG	Professional MPEG Forum
PS	Program Stream
PSI	Program Specific Information
PSTN	Public Switched Telephone Network
RDF	Resource Description Framework
RMP	Rights Management Protection
RMPS	Rights Management Protection System
RPC	Remote Procedure Call
RS	Reed–Solomon (RS)
RTP	Real-Time Transport Protocol
RTSP	Real-Time Streaming Protocol
S2S	Service Provider-to-Service Provider
SDTV	Standard Definition Television
SE	Service Editors
SGML	Standard Generalized Markup Language
SMPTE	Society of Motion Picture and Television Engineers
SMS	Short Message
SMTP	Simple Mail Transfer Protocol
SOAP	Simple Object Access Protocol
SP	Service Providers
STB	Set-Top Box
TBC	Tree Branch Code
TCP	Transport Control Protocol
TCP/IP	Transmission Control Protocol/Internet Protocol
TS	Transport Stream

UDDI	Universal Description, Discovery and Integration
UDP	User Datagram Protocol
UMA	Universal Multimedia Access
UMF	Unified Material Format
UU	User–User
VM	Virtual Machine
VoD	Video–on–Demand
VR	Virtual Reality
VRML	Virtual Reality Modeling Language
W3	World Wide Web Consortium
WSDL	Web Services Description Language
WWW	World Wide Web
XML	eXtensible Markup Language

List of Figures

List of Tables

References

[1] ISO/IEC 13818-1. Generic coding of moving pictures and associated audio — part 1: Systems, 1996. Recommendation H.222.0.

[2] ISO/IEC 15938-5. *Information Technology - Multimedia Content Description Interface—part 5 Multimedia Description Schemes.* International Organisation for Standardisation (ISO/IEC), 2001.

[3] EN 301 192. *DVB: Specification for Data Broadcasting.* European Telecommunications Standards Institute (ETSI), December 1997.

[4] ISO/IEC JTC 1/SC29/WG11. *Information Technology — Multimedia Content Description Interface.* International Organisation for Standardisation (ISO/IEC), April 2002.

[5] ETS 300 468. *DVB: Specification for Service Information (SI) in DVB Systems.* European Telecommunications Standards Institute (ETSI), January 1997.

[6] ETS 300 800. *DVB: Interaction Channel for Cable TV Distribution Systems.* European Telecommunications Standards Institute (ETSI), Jannuary 1997.

[7] ETS 300 801. *DVB: Interaction Channel through Public Switched Telecommunications Network (PSTN)/Integrated Services Digital Networks (ISDN).* European Telecommunications Standards Institute (ETSI), August 1997.

[8] ETS 300 802. *DVB: Network-Independend Protocols for DVB Interactive Systems.* European Telecommunications Standards Institute (ETSI), November 1997.

[9] ISO/IEC 8879. *Information Processing—Text and Office Systems—Standard General MarkUp Language (SGML).* International Organisation for Standardisation (ISO/IEC), 1986.

[10] Advanced Authoring Format. IBC 2001 slide-set presentation, 2001.

[11] J.L. Abraham and D.J. Knight. Strategic innovation: Leveraging creative action for more profitable growth. *IEEE Engineering Management Review*, 30(4):23–28, 2002.

[12] A.V. Aho, R. Sethi, and J. D. Ullman. *Compilers. Principles, Techniques, and Tools.* Addison-Wesley, Reading, MA, 2nd edition, 1986.

[13] D. Airola, L. Boch, and G. Dimino. Automated ingestion of audiovisual content. In *IBC 2002 Proceedings*, 2002.

[14] AAF Association. Advanced Authoring Format (AAF) Association. www.aafassociation.org.

[15] AAF Association. Enabling better media workflows. www.aafassociation.org, September 2001.

[16] Audio-Video Transport Working Group, H. Schulzrinne, S. Casner, R. Frederick, and V. Jacobson. RFC 1889: RTP: A transport protocol for real-time applications, January 1996. Status: PROPOSED STANDARD.

[17] O. Avaro and P. Salembier. MPEG-7 systems: Overview. *IEEE Transactions on Circuits and Systems for Video Technology*, 11(6), June 2001.

[18] P. Baldi and S. Brunak. *Bioinformatics.* MIT Press, Cambridge, MA, 1998.

[19] Ned Batchelder. Metadata is nothing new. Internet: `http://www.nedbatchelder.com/`.

[20] BBC. BBC standard media exchange framework (SMEF). http://www.bbc.co.uk/guidelines/smef.

[21] D. Beenham, P. Schmidt, and G. Sylvester-Bradley. XML based dictionaries for MXF/AAF applications. In *IBC 2002 Proceedings*, 2002.

[22] H. Benoit. *Digital Television: MPEG-1, MPEG-2 and Principles of the DVB system.* Arnold, Copublished with John Wiley & Sons, London, UK, 1997.

[23] T. Berners-Lee, R. Fielding, and L. Masinter. RFC 2396: Uniform Resource Identifiers (URI): Generic syntax, August 1998. Status: DRAFT STANDARD.

[24] E. Bertino and E. Ferrari. Temporal synchronization models for multimedia data. *IEEE Transactions on Knowledge and Data Engineering*, 10(4), July–August 1998.

[25] D. Bordwell and K. Thompson. *Film Art: An Introduction.* McGraw-Hill, New York, 5th edition, 1997.

[26] J. Bormans, J. Gelissen, and A. Perkis. MPEG-21: The 21st century multimedia framework. *IEEE Signal Processing Magazine*, 20(2):53–62, March 2003.

[27] L. Böszörményi, M. Döller, H. Hellwagner, H. Kosch, M. Libsie, and P. Schojer. Comprehensive treatment of adaptation in distributed multimedia systems in the ADMITS project. In *Proceedings of the Tenth ACM International Conference on Multimedia (MM-02)*, pages 429–430, New York, December 2002. ACM Press.

[28] V.M. Bove, J. Dakss, E. Chalom, and S. Agamanolis. Hyperlinked television research at the MIT media laboratory. *IBM Systems Journal*, 39(3 and 4), 2000. `http://www.research.ibm.com/journal/sj/393/part1/bove.html`.

[29] R. Braden, Ed., L. Zhang, S. Berson, S. Herzog, and S. Jamin. RFC 2205: Resource ReSerVation Protocol (RSVP) — version 1 functional specification, September 1997. Status: PROPOSED STANDARD.

[30] T. Bray, J. Paoli, and C.M. Sperberg-McQueen, editors. *Extensible Markup Language (XML) 1.0.* Sp, February 1998.

[31] S. Brinkman and S. Flank. Drinking from the fire hose: Managing metadata. In *IBC 2002 Proceedings*, 2002.

[32] C. Brotherton. *Social Psychology and Management.* Milton Keynes: Open University Press, Bristol, PA, 1999.

[33] CENELEC. EN 50221: common interface for conditional access and other DVB decoder applications, 1997.

[34] D.E. Comer. *Internetworking with TCP/IP: Vol. I: Principles, Protocols, and Architecture.* Prentice-Hall, Upper Saddle River, New Jersey, 3rd. edition, 1995.

[35] European Commission. Directive 95/47/EC of the European Parliament and the Council of 24th October 1995 on the use of standards for the transmission of television signals. Directive, October 1995.

[36] H. Comon, M. Dauchet, R. Gilleron, F. Jacquemard, D. Lugiez, S. Tison, and M. Tommasi. Tree automata techniques and applications. This electronic book is available at `http://www.grappa.univ-lille3.fr/tata`, 1999.

[37] World Wide Web Consortium. HyperText Markup Language Specification—2.0. Internet Draft, February 1995. Expires June 19, 1995.

[38] World Wide Web Consortium. XSL Transformations (XSLT). W3C Recommendation, 1999. `http://www.w3.org/TR/xslt`.

[39] World Wide Web Consortium. Extensible markup language (XML) 1.0 (2nd edition) — W3C recommendation. Available at http://www.w3.org/TR/2000/WD-xml-2e-20000814, 2000.

[40] World Wide Web Consortium. XML path language (XPath) version 1.0—W3C recommendation. Available at http://www.w3.org/TR/xpath.html, 2000.

[41] World Wide Web Consortium. Scalable Vector Graphics (SVG)—1.0. W3C Recommendation, September 2001.

[42] World Wide Web Consortium. XML Schema Part 0: Primer. W3C Recommendation, May 2001.

[43] World Wide Web Consortium. XML Schema Part 2: Datatypes. W3C Recommendation, May 2001.

[44] World Wide Web Consortium. XQuery: The W3C query language for XML—W3C working draft. Available at http://www.w3.org/TR/xquery/, 2001.

[45] D. J. Cutts. DVB conditional access. In *Broadcasting Convention, International (Conf. Publ. No. 428)*, 1996.

[46] M. Day. Metadata in a nutshell. *Information Europe*, 6(2), 2001. Information Europe is published by EBLIDA (the European Bureau

of Library, Information and Documentation Associations) and this article can be found onlin on `http://www.ukoln.ac.uk/metadata/publications/nutshell/`.

[47] Dublin Core Metadata Intitiative (DCMI). Dublin core metadata initiative (dcmi). www.dublincore.org.

[48] J. Delgado. Standardization of the management of intellectual property rights in multimedia content. In *Proceedings of the Second International Conference on Web Delivery of Music (WEDELMUSIC)*, 2002.

[49] J. Delgado, I. Gallego, and E. Rodriguez. Use of the MPEG-21 rights expression language for music distribution. In *Proceedings of the Third International Conference on Web Delivery of Music (WEDELMUSIC)*, 2003.

[50] S. DeRose, E. Maler, and D. Orchard. XML Linking Language (XLink) version 1.0—W3C proposed recommendation 20 December 2000. Technical Report PR-xlink-20001220, World Wide Web Consortium, December 2000.

[51] S.J. DeRose, E. Maler, and R. Daniel (Eds). XML Pointer Language (XPointer) Version 1.0. W3C Last Call Working Draft, January 2001. http://www.w3.org/TR/xptr.

[52] B. Devlin. MXF the Material eXchange Format. *EBU Technical Review*, July 2002.

[53] MOT Dictionary. http://www.kielikone.fi/en/.

[54] T. Dierks and C. Allen. RFC 2246: The TLS protocol version 1, January 1999. Status: PROPOSED STANDARD.

[55] F. D'Souza, Desmond and Cameron Willis, Alan. *Objects, Components, and Frameworks with UML*. Addison Wesley, Reading, MA, 1999.

[56] Digital Video Broadcasting (DVB). Digital video broadcasting (DVB). `http://www.dvb.org`.

[57] European Broadcasting Union (EBU). www.ebu.ch.

[58] European Broadcasting Union (EBU). EBU Tech 3295: The EBU metadata exchange scheme. Technical report, European Broadcasting Union (EBU), 2003.

[59] B. Edge. GXF the general exchange format. *EBU Technical Review*, July 2002.

[60] Francisco Curbera et al. Unraveling the Web services web: An introduction to SOAP, WSDL, and UDDI. *IEEE Distributed Systems Online*, 3(4), 2002.

[61] K. Duffey et al. *Professional JSP Site Design: Coding Core Web Applications*. Wrox, Hoboken, NJ, 2001.

[62] S. Brown et al. *Professional JSP*. Wrox Press, Hoboken, NJ, 2nd edition, 2001.

[63] European Telecommunications Standards Institute (ETSI). EBU project group P/FRA. `http://www.ebu.ch/pmc_fra.html`.

[64] European Telecommunications Standards Institute (ETSI). Technical report ETR289: Support for the use of scrambling and conditional access (ca) within digital broadcasting systems, October 1996.

[65] European Telecommunications Standards Institute (ETSI). *EN200472: Digital Video Broadcasting (DVB): Specification for conveying ITU-R System B Teletext in DVB bitstreams*, August 1997.

[66] European Telecommunications Standards Institute (ETSI). ETS300743: Digital Video Broadcasting (DVB): Subtitling systems, 1997.

[67] European Telecommunications Standards Institute (ETSI). TR10194: Digital Video Broadcasting (DVB), guidelines for implementation and usage of the specification of network independent protocols for DVB interactive services, June 1997.

[68] European Telecommunications Standards Institute (ETSI). *Digital Video Broadcasting (DVB): Multimedia Home Platform (MHP)*, October 2001.

[69] U. Fayyad, G. Piatetsky-Shapiro, and P. Smyth. The KDD process for extracting useful knowledge from volumes of data. *Communications of the ACM*, 39(11):27–34, 1996.

[70] U. Fayyad, G. Piatetsky-Shapiro, P. Smyth, and R. Uthurusamy. *Advances in Knowledge Discovery and Data Mining.* AAAI Press/MIT Press, Cambridge, MA, 1996.

[71] R. Fielding, J. Gettys, J. Mogul, H. Frystyk, and T. Berners-Lee. RFC 2068: Hypertext Transfer Protocol—HTTP/1.1, January 1997. Status: PROPOSED STANDARD.

[72] B. Foote. DVB's gem: Bringing MHP's sparkle to the US and byond. *DVB Scene 05*, 2003. `http://www.dvb.org`.

[73] EBU/SMPTE Task Force for Harmonized Standards for the Exchange of Programme Material as Bitstreams. Final report: Analyses and results. Technical report, EBU/SMPTE, 1998.

[74] Society for Motion Picture and Television Engineers. www.smpte.org.

[75] Society for Motion Picture and Television Engineers. SMPTE 300M-2000: Unique material identifier (UMID). SMPTE Standard, 2000.

[76] Society for Motion Picture and Television Engineers. SMPTE 335M-2001: Metadata dictionary structure. SMPTE Standard, 2001.

[77] Society for Motion Picture and Television Engineers. SMPTE 336M-2001: Data encoding protocol using key-length-value (KLV). SMPTE Standard, 2001.

[78] Society for Motion Picture and Television Engineers. SMPTE 360M-2001: General exchange format (GXF). SMPTE Standard, 2001.

[79] Society for Motion Picture and Television Engineers. SMPTE RP210.4-2002: Metadata dictionary registry of metadata element descriptions. SMPTE Recommended Practice, 2002.

[80] Pro-MPEG Forum. Asset management: Working together with MXF. `http://www.broadcastpapers.com/asset/ProMPEGForumMXF01.htm`.

[81] TV-Anytime Forum. TV Anytime Forum. http://www.tv-anytime.org, 1999.

[82] TV-Anytime Forum. R5 rights management and protection requirements. TV-Anytime Requirements, 2000.

[83] TV-Anytime Forum. S5 rights management and protection specification, working draft. TV-Anytime Working Draft, 2002.

[84] TV-Anytime Forum. S6 metadata services over a bi-directional network v1.0. TV-Anytime Specification, 2002.

[85] TV-Anytime Forum. S7 bi-directional metadata delivery protection v1.0. TV-Anytime Specification, 2002.

[86] TV-Anytime Forum. S1 phase 1 benchmark features (informative) v1.2. TV-Anytime Specification, 2003.

[87] TV-Anytime Forum. S2 system description (informative with mandatory appendix b) v1.3. TV-Anytime Specification, 2003.

[88] TV-Anytime Forum. S3 metadata (normative) v1.3. TV-Anytime Specification, 2003.

[89] TV-Anytime Forum. S4 content referencing (normative) v1.2. TV-Anytime Specification, 2003.

[90] K. Gajos, H. Fox, and H. Shrobe. End user empowerment in human centered pervasive computing. In *Short Paper Track at Pervasive 2002*, Zurich, Swizerland, 2002.

[91] B. Gilmer. Ingest, data and metadata. slide show, 2000.

[92] B. Gilmer. AAF the advanced authoring format. *EBU Technical Review*, July 2002.

[93] H. Gomaa. *Designing Concurrent, Distributed, and Real-Time Applications with UML.* Addison Wesley, Reading, MA, 2000.

[94] A. Greimers and J. Courtes. *Semiotics and Language: An Analytical Dictionary.* Indiana University Press, Bloomington, IN, 1982.

[95] IST Advisory Group. Ambient intelligence: from vision to reality. Draft report, European Union, IST Advisory Group, 2003.

[96] A. Grove. *Only the Paranoid Survive.* Doubleday: A Currency Book, New York, 1996.

[97] A. Holzinger, T. Kleinberger, and P. Müller. Multimedia learning systems based on IEEE learning object metadata (LOM). In *World Conference on Educational Multimedia: Hypermedia and Telecommunications*, volume 2001, pages 772–777, 2001.

[98] J.E. Hopcroft, R. Motwani, and J. D. Ullman. *Introduction to Automata Theory, Languages, and Computation.* Addison-Wesley, Reading, MA, 2nd edition, 2000.

[99] R. Hopper. EBU project group P/META metadata exchange standards. *EBU Technical Review*, September 2000.

[100] R. Hopper. EBU project group P/META—metadata exchange scheme, v1.0. *EBU Technical Review*, April 2002.

[101] P. Hoschka. Television and the Web. Internet (Talk, W3C), 1999. http://www.w3.org/Talks/1999/0512-tvweb-www8/.

[102] J. Ibbotson. XML protocol usage scenarios. http://www.w3.org/2000/xp/Group/1/11/19/UsageScenarios, 2000.

[103] ISO/IEC 21000-1. Multimedia framework—part 1: Vision, technologies and strategy. http://www.iso.ch/iso/en/ittf/PubliclyAvailableStandards, 2002.

[104] ISO/IEC 21000-10 (N5855). MPEG-21 multimedia framework—part 10: Digital item processing (working draft). http://www.iso.ch/iso/en/ittf/PubliclyAvailableStandards, 2003.

[105] ISO/IEC 21000-11 (N5875). MPEG-21 multimedia framework—part 11: Evaluation of persistent association tools (working draft). http://www.iso.ch/iso/en/ittf/PubliclyAvailableStandards, 2003.

[106] ISO/IEC 21000-12 (N5640). MPEG-21 multimedia framework—part 12: Resource delivery test bed (working draft). http://www.iso.ch/iso/en/ittf/PubliclyAvailableStandards, 2003.

[107] ISO/IEC 21000-2. Information technology-multimedia framework—part 2: Digital item declaration. http://www.iso.ch/iso/en/ittf/PubliclyAvailableStandards, 2003.

[108] ISO/IEC 21000-3. Information technology-multimedia framework—part 3: Digital item identification. http://www.iso.ch/iso/en/ittf/PubliclyAvailableStandards, 2003.

[109] ISO/IEC 21000-4. Multimedia framework—part 4: Intellectual property management and protection. Currently in progress, http://www.iso.ch/iso/en/ittf/PubliclyAvailableStandards.

[110] ISO/IEC 21000-5 (N5939). MPEG-21 multimedia framework—part 5: Rights expression language (REL)(fds). http://www.iso.ch/iso/en/ittf/PubliclyAvailableStandards, 2003.

[111] ISO/IEC 21000-6 (N5842). MPEG-21 multimedia framework—part 6: Rights data dictionary (RDD)(fds). http://www.iso.ch/iso/en/ittf/PubliclyAvailableStandards, 2003.

[112] ISO/IEC 21000-7 (N5845). MPEG-21 multimedia framework—part 7: Digital item adaptation (fcd). http://www.iso.ch/iso/en/ittf/PubliclyAvailableStandards, 2003.

[113] ISO/MPEG N4206. Multimedia content description interface—part 6 reference software. MPEG Systems Group, July 2001.

[114] ISO/MPEG N4224. Multimedia content description interface—part 4 audio. MPEG Systems Group, July 2001.

[115] ISO/MPEG N4242. Multimedia content description interface—part 5 multimedia description schemes. MPEG Systems Group, July 2001.

[116] ISO/MPEG N4285. Multimedia content description interface—part 1 systems. MPEG Systems Group, July 2001.

[117] ISO/MPEG N4288. Multimedia content description interface—part 2 description definition language. MPEG Systems Group, July 2001.

[118] ISO/MPEG N4358. Multimedia content description interface—part 3 visual. MPEG Systems Group, July 2001.

[119] T. Kleinberger, L. Schrepfer, A. Holzinger, and P. Müller. A multimedia repository for online educational content. In *World Conference on Educational Multimedia: Hypermedia and Telecommunications*, volume 2001, pages 975–980, 2001.

[120] E. Levinson. RFC 2111: Content-ID and message-ID uniform resource locators, February 1997. Obsoleted by RFC2392 [122]. Status: PROPOSED STANDARD.

[121] E. Levinson. RFC 2387: The MIME multipart/related content-type, August 1998.

[122] E. Levinson. RFC 2392: Content-ID and message-ID uniform resource locators, August 1998.

[123] A. Lugmayr and S. Kalli. Taxonomy of XML based metadata in a real-time digiTV deployment environment: Digital broadcast item taxonomy. In *Real-Time Imaging 2003*. SPIE, 2003. accepted, not published.

[124] A. Lugmayr and S. Kalli. Metadata-based svg and x3d graphical in interactive tv. In V. Geroimenko and C. Chen, editors, *Visualizing Information Using SVG and X3D*. Springer Verlag, New York, to be published in 2004.

[125] A. Lugmayr, S. Kalli, and R. Creutzburg. Synchronization of MPEG-7 metadata with a broadband MPEG-2 digiTV stream by utilizing a digital broadcast item approach. In *Low-Light-Level and Real-Time Imaging Systems, Components, and Applications*, pages 207–217, Washington, USA, 2003. SPIE.

[126] A. Lugmayr, J. Lyytinen, A. Mailaparampil, F. Tico, S. Maijala, S. Tuominen, T. Pihlajamäki, P. Rautavirta, M. Oksanen, J. Spieker, S. Niiranen, and S. Kalli. The future interaction TV project developing DIET—digital interaction environment for TV. In *Proceedings of the fifth Nordic Signal Processing Symposium: NORSIG 2002*. NORSIG, October 2002.

[127] A. Lugmayr, A. Mailaparampil, F. Tico, S. Kalli, and R. Creutzburg. A digital broadcast item (dbi) enabling metadata repository for digital interactive television (digiTV) feedback channel networks. In *SPIE-EI 2003*, Washington, USA, 2003. accepted.

[128] A. Lugmayr, S. Niiranen, A. Mailaparampil, P. Rautavirta, M. Oksanen, F. Tico, and S. Kalli. Applying MPEG-21 in digital television—example use scenarios: e-postcard, e-game, and e-ticket. In *IEEE International Conference on Multimedia and Expo*, 2002.

[129] L.A. Maciaszek. *Requirements Analysis and System Design*. Addison Wesley, Reading, MA, 2001.

[130] S. Manjunath, B., Phillippe Salembier, and Thomas Sikora, editors. *Introduction to MPEG-7*. Wiley, 2002.

[131] Merriam-Webster. *Merriam-Webster Dictionary*. Merriam-Webster, online. `http://www.m-w.com/cgi-bin/dictionary`.

[132] M. Milenkovic. Delivering interactive services via a digital tv infrastructure. *IEEE Multimedia*, 5(4):34–43, October–December 1998.

[133] G. Mills. DVB 2.0: Making progress. *DVB Scene 06*, 2003. http://www.dvb.org.

[134] O. Morgan. An introduction to the advanced authoring format. Oral presentation, 2000.

[135] M. Murata, D. Lee, and M. Mani. Taxonomy of XML schema languages using formal language theory. *Extreme Markup Languages 2000*, 2000.

[136] D.R. Myers, C.W. Sumpter, S.T. Walsh, and B.A. Kirchhoff. Guest editorial: A practitioner's view: Evolutionary stages of disruptive technologies. *IEEE Transactions on Engineering Management*, 49(4):322–329, November 2002.

[137] J. Myers and M. Rose. RFC 1939: Post Office Protocol—version 3, May 1996.

[138] N. Negroponte. Object-oriented television. Internet, 1996. http://web.media.mit.edu/~nicholas/Wired/WIRED4-07.html.

[139] R. Nelson. RFC 1957: Some observations on implementations of the Post Office Protocol (POP3), June 1996. Updates RFC1939 [137]. Status: INFORMATIONAL.

[140] J.C. Newell. An introduction to MHP 1.0 and MHP 1.1, May 2002. R&D white paper.

[141] J. Niinimaeki, A. Holopainen, J. Kerttula, and J. Reponen. Security development of a pocket-sized teleradiology consultation system. In *Medinfo 2001*, volume 10, 2001.

[142] S. Niiranen. Broadcast multimedia personalization. Licentiate thesis, Tampere University of Technology, 2003.

[143] S. Niiranen, A. Lugmayr, H. Lamminen, and S. Kalli. Security in the health care applications of digital television. In *Proceedings of the Fifth Nordic Signal Processing Symposium: NORSIG 2002*. NORSIG, October 2002.

[144] National Institute of Standards and Technology (NIST). Pervasive computing program. http://www.itl.nist.gov/pervasivecomputing.html, 2001.

[145] Z. Pawlak, J. Grzymala-Busse, R. Slowinski, and W. Ziarko. Rough sets. *Communications of the ACM*, 38(11):89–95, 1995.

[146] M. Porter. *Competitive Strategy: Techniques for Analyzing Industries and Competitors*. The Free Press, New York, 1980.

[147] J. Postel. RFC 768: User datagram protocol, August 1980.

[148] J. Postel. RFC 791: Internet protocol, September 1981.

[149] J. Postel. RFC 1591: Domain name system structure and delegation, March 1994. Status: INFORMATIONAL.

[150] Pro-MPEG-Forum. Material Exchange Format (MXF). www.pro-mpeg.org.

[151] D. J. Rayers. Metadata in TV production: Associating the tv production processes with relevant technologies. In *IBC 2002 Conference Publication*, Amsterdam, September 2002. IBC.

[152] M. Reid and R. Hammersley. *Communicating Successfully in Groups.* Routledge, New York, 2000.

[153] P. Riceur. Recherches en communication no. 7, le recit mediatique, 1997.

[154] M.L. Ryan. *Possible Worlds, Artificial Intelligence and Narrative Theory.* Indiana University Press, Bloomington, IN, 1991.

[155] M.L. Ryan. *Cyberspace Textuality: Computer Technology and Literary Theory.* Indiana University Press, Bloomington, IN, 1999.

[156] M.L. Ryan. *Narrative as Virtual Reality: Immersion and Interactivity in Literature and Electronic Media.* Johns Hopkins University Press, Baltimore, MD, 2001.

[157] J. Samsel and D. Wimberley. *Writing for Interactive Media.* Allworth, New York, 1998.

[158] H. Schulzrinne, A. Rao, and R. Lanphier. RFC 2326: Real time streaming protocol (RTSP), April 1998. Status: PROPOSED STANDARD.

[159] D. Sitaram and A. Dan. *Multimedia Servers.* Morgan Kaufmann, San Mateo, CA, 2000.

[160] SOAP. *Simple Object Access Protocol (SOAP 1.1).* World Wide Web Consortium (W3C), May 2000.

[161] J. Song, S.J. Yang, C. Kim, J. Nam, J.W. Hong, and Y.M. Ro. Digital item adaptation for color vision variations. In *Proceedings of IS&T and SPIEI S&T/SPIE's Fifteenth Annual Symposium: Human Vision and Electronic Imaging VIII, Electronic Imaging Science and Technology,* 2003. accepted but not published.

[162] H. Soronen. Tulevaisuuden sähköiset palvelut eri päätelaitteissa, digitalisoituvan viestinnän monet kasvot. Teknologiakatsaus 118/2001, TEKES, 2001.

[163] H. Sun, A. Verro, and K. Asai. Resource adaptation based on MPEG-21 usage environment descriptions. In *Proceedings of the 2003 International Symposium on Circuits and Systems ISCAS '03*, volume 2, pages 536–539, May 2003.

[164] M. Takahashi. Generalizations of regular sets and their application to a study of context-free languages. *Information and Control*, 27(1), January 1975.

[165] Tektronix. *A Guide to MPEG Fundamentals and Protocol Analysis.* Tektronix Inc., 1997.

[166] H. S. Thompson, D. Beech, M. Maloney, and N. Mendelsohn (Eds). Xml schema part 1: Structures. W3C Recommendation, May 2001. http://www.w3.org/TR/xmlschema-1/.

[167] Understanding the AAF. DV Production Workshop NAB 2000, slide-set presentation, April 2000.

[168] World Wide Web Consortium (W3C). XML schema Part 1: Structures. W3C Recommendation, May 2001.

[169] P. Wallich. Digital hubbub. *IEEE Spectrum*, 39(7):26–31, July 2002.

[170] WDSL. *Web Services Description Language (WSDL 1.1).* World Wide Web Consortium (W3C), May 2001.

[171] M. Weiser. Hot topics—ubiquitous computing. *IEEE Computer*, 26(10):71–72, October 1993.

[172] World Wide Web Consortium (W3C). *SOAP Messages with Attachments*, December 2000.

[173] T. Worthington. Metadata: The killer application for digital broadcasting? `http://www.tomw.net.au/2002/mka.html`, 2002.

[174] Y. Zhang. Virtual communities and team formation. In *Crossroads—The ACM Student Magazine*, 2003. Fall 2003, 10.1.

[175] W. Ziarko. Discovery through rough set theory. *Communications of the ACM*, 42(11):55–57, November 1999.

Index

Fig. 2.10 Shopping in digital TV may involve both professional eShops and services for consumers to sell private goods (© Ortikon Interactive Ltd.)

Fig. 2.14 Online community service for young people (© Zento Interactive Ltd.)

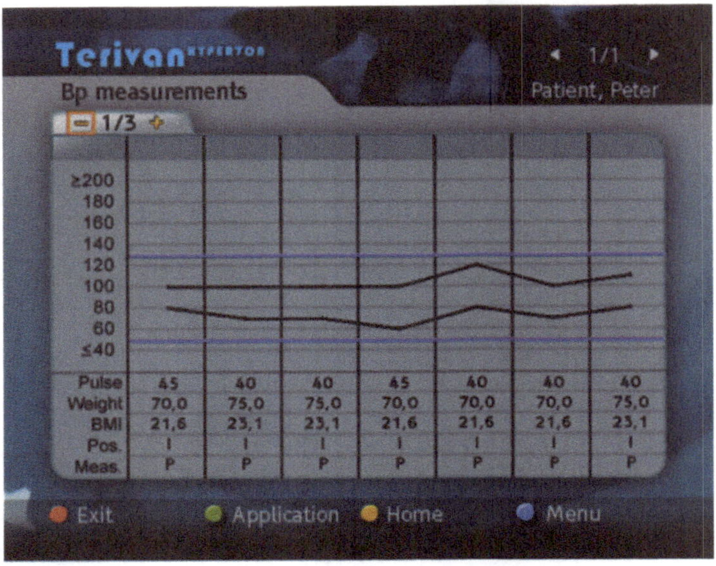

Fig. 2.15 Blood pressure monitoring in digital TV (© Terivan Ltd.)

Fig. 6.5 "Habbo Hotel" creates 3-D chatting spaces (© Sulake Ltd.)

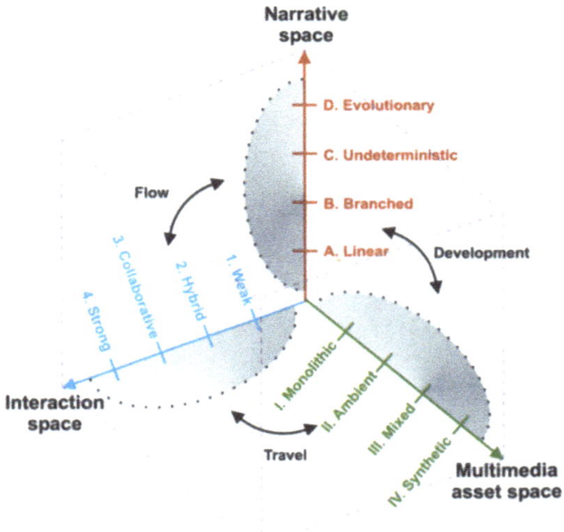

Fig. 6.9 Fictive digital TV universe — the narrative cube

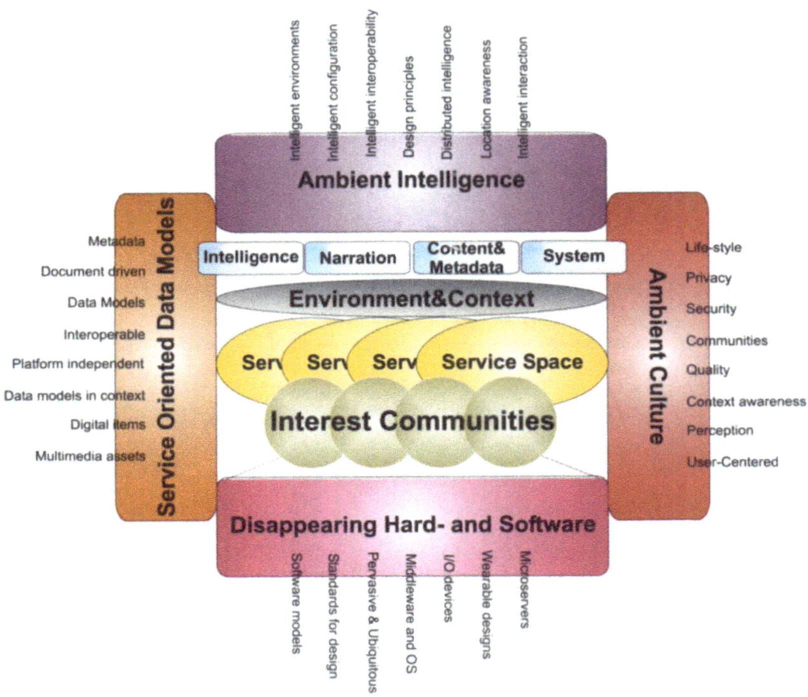

Fig. 11.1 The environment as "browser". Ambient TV as concept

Fig. 12.1 Novel screens displaying images appearing to float in the air (© Fog Screen Inc.)

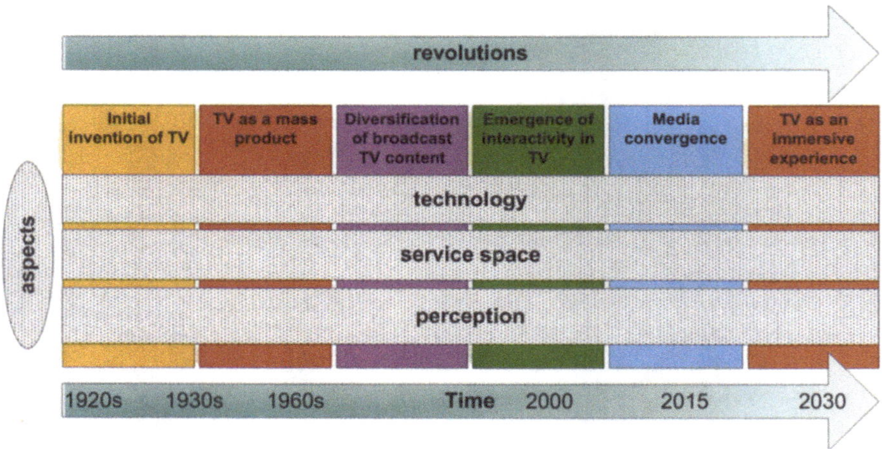

Fig. 13.1 Factors in broadcast multimedia and their evolution

Kolibri 360 GmbH
Ferdinand-Jühlke-Straße 7
99095 Erfurt, Deutschland
produktsicherheit@kolibri360.de